Basic Electricity

REVISED EDITION

VAN VALKENBURGH,
NOOGER & NEVILLE, INC.

VOL. 1

HAYDEN BOOK COMPANY, INC.
Rochelle Park, New Jersey

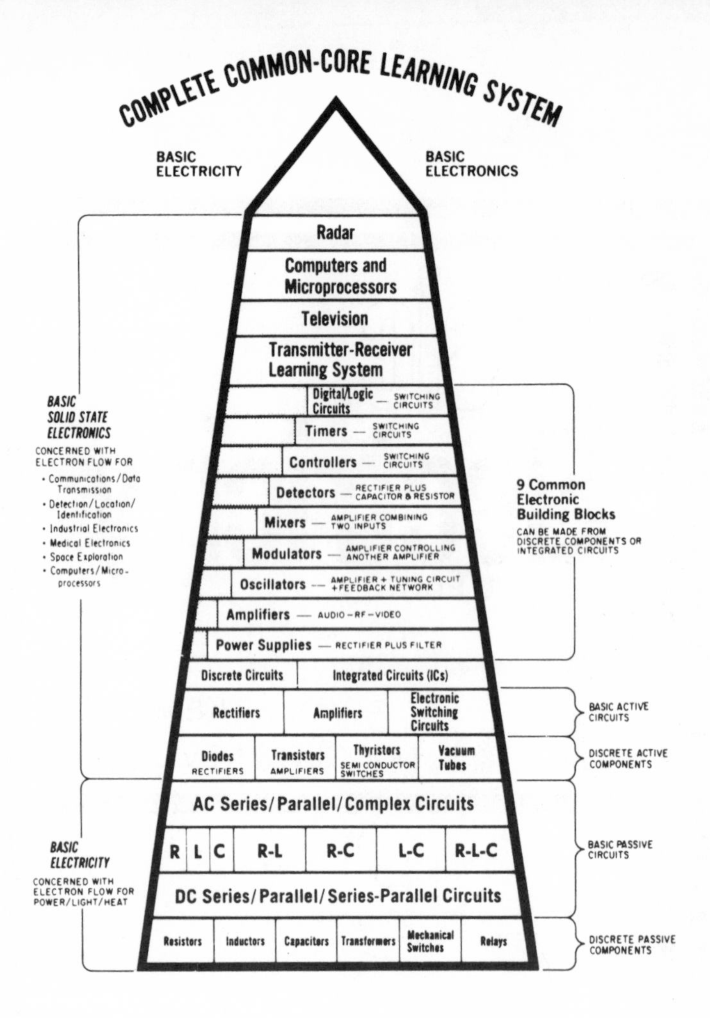

Fifth Printing — 1981

ISBN 0-8104-0876-7
Library of Congress Catalog Card Number 76-57841

Printed in the United States of America

Preface to Revision of BASIC ELECTRICITY

The COMMON-CORE® Program – *Basic Electricity, Basic Electronics, Basic Synchros and Servomechanisms,* etc. – was designed and developed during the years 1952–1954. On the basis of a job task analysis of a broad spectrum of U.S. Navy electrical/electronics equipment of that era, there was established a "common-core" of prerequisite knowledge and skills. This "common-core" prerequisite was then programed into a teaching/learning system which had as its primary instructional objective the effective training of U.S. Navy electrical/electronics technicians who could understand and apply such understanding in meaningful job problem situations.

Since that time, over 100,000 U.S. Navy technicians have been efficiently trained by this performance-based system. Civilian students and technicians have accounted for hundreds of thousands more. The military and civilian education and training programs in South America, Europe, the Middle East, Asia, Australia, and Africa have also recognized its usefulness with some 12 foreign-language editions presently in print.

Now the foundation of the COMMON-CORE Program, *Basic Electricity,* is being updated and improved. Its equipment job task base has been enlarged to cover the understanding and skills needed for the spectrum of present-day electrical/electronic equipment – modern industrial machines, controls, instrumentation, computers, communications, radar, lasers, etc. Its technological components/circuits/functions base has been revised and broadened to incorporate the generations of development in electrical/electronics technology – namely, from (1) vacuum tubes to (2) transistors and semiconductors to (3) integrated circuits, large scale integration, and microminiaturization.

Educationally, considerable effort has been given to incorporating individualized learning/testing features and techniques within the texts themselves, and in the accompanying interactive student mastery tests.

Notwithstanding the passage of time, the original innovative, basic text-format, system-design elements of the COMMON-CORE Program still stand – this solid framework of proved effectiveness that has been the stimulus for many of the improvements in vocational/technical education.

VAN VALKENBURGH, NOOGER AND NEVILLE, INC.

New York, N.Y.

CONTENTS

How Current Is Measured

How Voltage Is Measured

What Controls Current Flow—Resistance

Review

Introducing Ohm's Law

SPACE
INDUSTRY
ELECTRICITY
IS THE KEY TO
OPPORTUNITY
IN...
SCIENCE
DEFENSE

The Importance of the Study of Electricity

It is hard to imagine a world without electricity. It touches and influences our daily lives in hundreds of ways. We see the use of electricity directly in our homes for lighting, the operation of appliances, telephone, television, radio, stereo, heating, etc. We see the use of electricity in transportation. Electricity has been used in the manufacture of most of the things we use either directly or to operate machines that make or process the products that we need. Without electricity, most of the things we use and enjoy today would not be possible.

Early History

The word electricity comes from the ancient Greek word for amber —*elektron*. The early Greeks observed that when amber (a fossilized resin) was rubbed with a cloth, it would attract bits of material such as dried leaves. Later, scientists showed that this property of attraction occurred in other materials such as rubber and glass but did not occur with materials such as copper or iron. The materials that had this property of attraction when rubbed with a cloth were described as being charged with an *electric force*; and it was noticed that some of these charged materials were attracted by a charged piece of glass and that others were repelled. Benjamin Franklin called these two kinds of charges (or electricity) *positive* and *negative*. We know now, as you will learn, that what was actually being observed was an excess or deficiency in the materials of particles called *electrons*.

From time to time various scientists found that electricity seemed to behave in a constant and predictable way in a given situation. These scientists described this behavior in the form of rules or laws. These laws allow us to predict how electricity will behave even though today we still do not know its precise nature. By learning the rules or laws applying to the behavior of electricity, and by learning the methods of producing and controlling and using it, you will have learned electricity.

The Electron Theory

All the effects of electricity take place because of the existence of a tiny particle called the *electron*. Since no one has actually seen an electron, but only the effects it produces, we call the laws governing its behavior the *electron theory*. The electron theory is not only the basis for the design for all electrical and electronic equipment, it also explains physical and chemical action and helps scientists to probe into the very nature of the universe and life itself.

The Electron Theory (continued)

Since assuming that the electron exists has led to so many important discoveries in electricity, electronics, chemistry, and atomic physics, we can safely assume that the electron really exists. All electrical and electronic equipment has been designed using this theory. Since the electron theory has always worked for everyone, it will always work for you.

Your entire study of electricity will be based upon the electron theory, which assumes that all electrical and electronic effects are due to the movement of electrons from place to place or that there are too many or too few electrons in a particular place.

According to the electron theory, all electrical and electronic effects are caused either by the movement of electrons from place to place or because there exist too many or too few electrons in a particular place at a particular time.

Before you can usefully begin to consider the forces that make electrons move or accumulate, you must first find out what an electron is.

All matter is composed of atoms of many different sizes, degrees of structural complexity, and weight. But all atoms are alike in consisting of a nucleus—which differs from atom to atom of the 100-odd chemical elements that either exist in Nature or have been made by man—and of a varying number of electrons which move about that nucleus.

You will get an idea of what the atom is essentially like by looking at the picture below.

THE ELECTRON IS ELECTRICITY

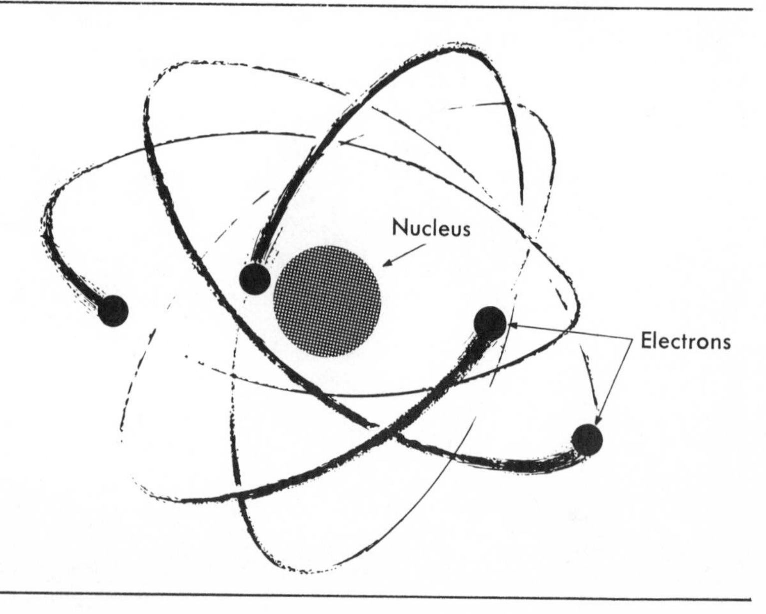

The Breakdown of Matter

A good way of understanding more about what the electron is like is to examine closely the composition of a drop of ordinary water.

If you take this drop of water and divide it into two drops, divide one of these two drops into two smaller drops, and repeat this process thousands of times, you will have a very tiny drop of water. This tiny drop will be so small that you will need the best microscope made today in order to see it.

This tiny drop of water will still have all the chemical characteristics of water. If examined by a chemist, he will not be able to find any chemical difference between this microscopic drop and an ordinary glass of water.

The Breakdown of Matter (continued)

Now if you take this tiny drop of water and try to divide it in half any further, you will not be able to see it in your microscope. Imagine then that you have available a super microscope that will magnify many times more than any microscope presently existing. This microscope can give you any magnification you want, so you can put your tiny drop of water under it and divide it into smaller and smaller droplets.

As the droplet of water is divided into smaller and smaller droplets, these tiny droplets will still have all the chemical characteristics of water. However, you eventually will have a droplet so small that any further division will cause it to lose the chemical characteristics of water. This last bit of water is called a *molecule.* Thus, a molecule is the smallest unit into which a substance can be divided and still be identified as that substance.

The Structure of the Molecule

When you increase the magnifying power of the microscope, you will see that the water molecule is made up of two tiny structures that are the same and a larger structure that is different from the two. These structures are called *atoms*. The two smaller atoms which are the same are hydrogen atoms and the larger, different one is an oxygen atom. When two atoms of hydrogen combine with one atom of oxygen, you have a molecule of water.

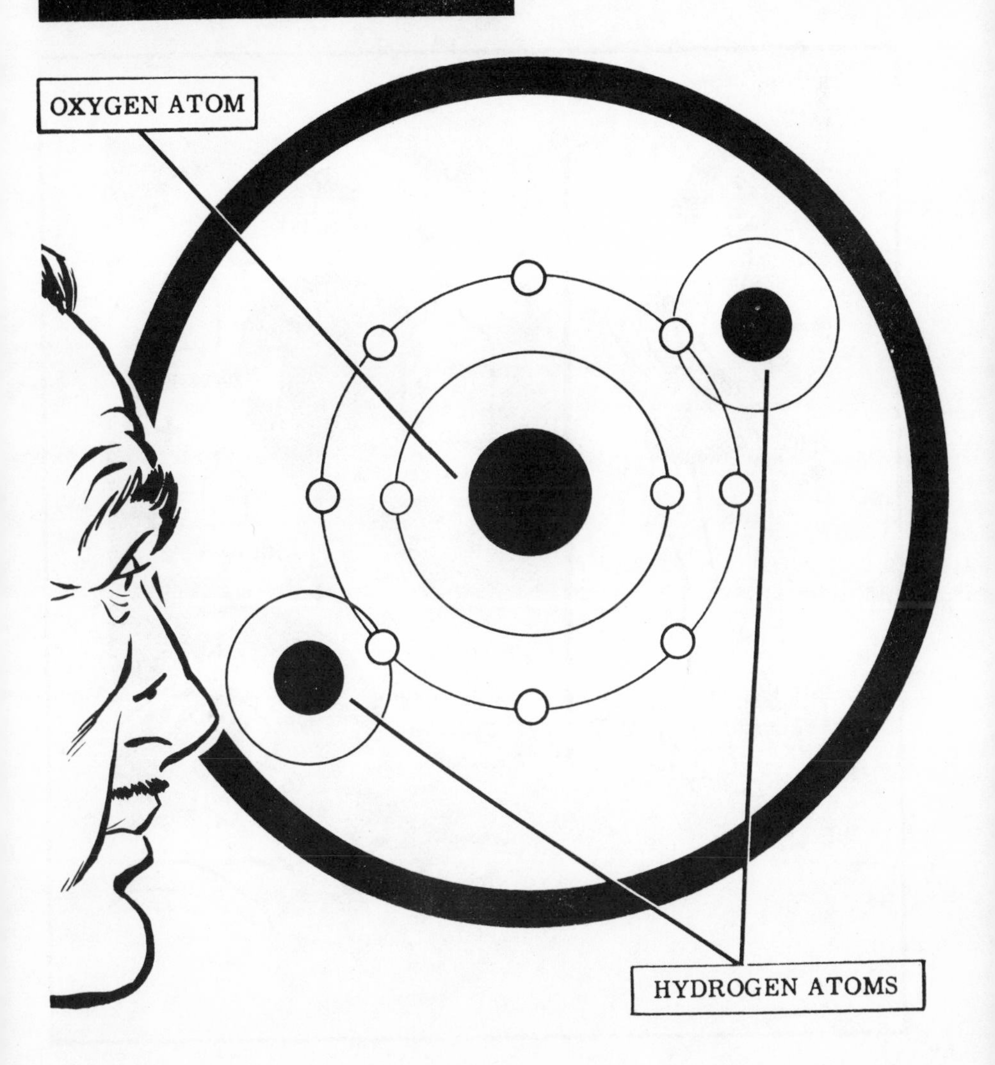

The Structure of the Molecule (continued)

While water is made up of only two kinds of atoms—oxygen and hydrogen—the molecules of many materials are more complex in structure. Cellulose molecules, the basic molecules of which wood is made, consist of three different kinds of atoms—carbon, hydrogen, and oxygen. All materials are made up of different combinations of atoms to form molecules of the materials. There are only about 100 different kinds of atoms and these are known as elements: oxygen, carbon, hydrogen, iron, gold, and nitrogen are all elements. The human body with all its complex tissues, bones, teeth, etc., is made up mainly of only 15 elements, and only six of these are found in quantity. (See Table of Elements at back of book.)

The Structure of the Atom

Now that you know that all materials are made up of molecules which consist of various combinations of about 100 different types of atoms, you will want to know what all this has to do with electricity. Increase the magnification of your imaginary super microscope still further and examine the atoms in the water molecule. Pick out the smallest atom you can see—the hydrogen atom—and examine it closely.

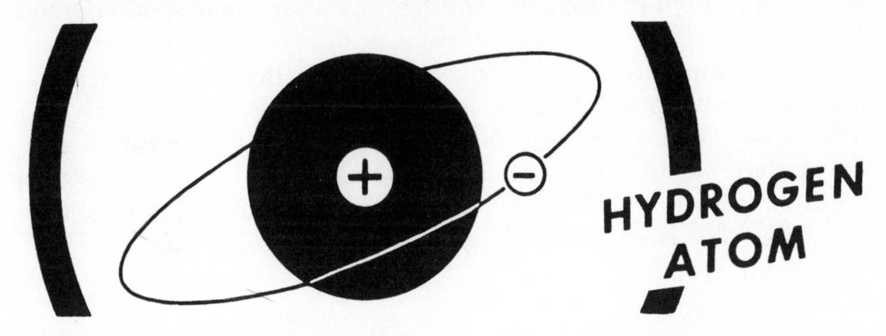

You see that the hydrogen atom is like a sun with one planet spinning around it. The planet is known as an electron and the sun is known as the nucleus. The electron has a *negative* (—) charge of electricity and the nucleus has a *positive* (+) charge of electricity.

In an atom, the total number of negatively charged electrons circling around the nucleus equals exactly the number of positive charges in the nucleus. The positive charges are called *protons.* Besides the protons, the nucleus also contains electrically neutral particles called *neutrons,* which are like a proton and an electron bonded together. Atoms of different elements contain different numbers of neutrons within the nucleus, but the number of electrons spinning about the nucleus always equals the number of protons (or positive charges) within the nucleus.

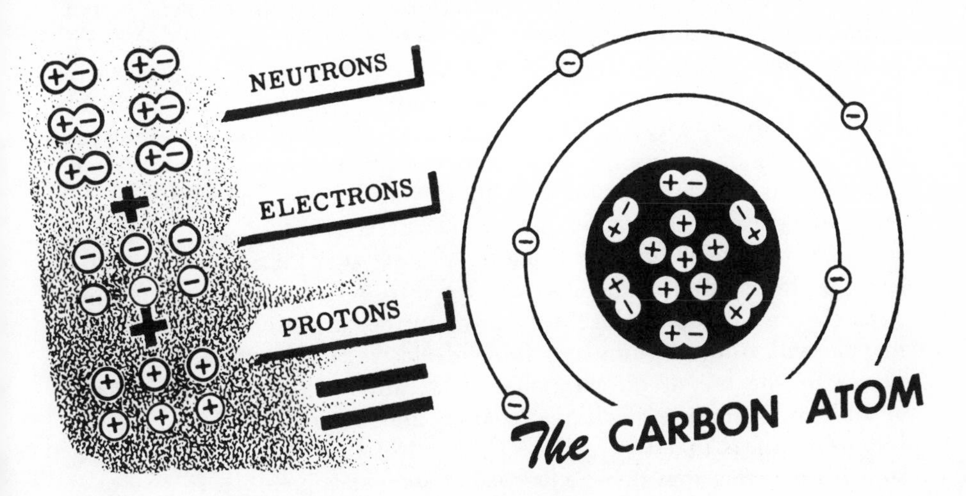

Electric Current and Electric Charge

All atoms are bound together by powerful forces of attraction between the nucleus and its electrons. Electrons in the outer orbits of an atom, however, are attracted to their nucleus less powerfully than are electrons whose orbits are nearer the nucleus.

In certain materials (they are known as electrical conductors), these outer electrons are so weakly bound to their nucleus that they can easily be forced away from it altogether and left to wander among other atoms at random.

Such electrons are called *free electrons*. It is the directional movement of free electrons which makes an electric current.

Electrons which have been forced out of their orbits create a deficiency of electrons in the atoms they leave and will cause a surplus of electrons at the point where they come to rest. A material with a deficiency of electrons is positively charged; one possessing a surplus of electrons is negatively charged.

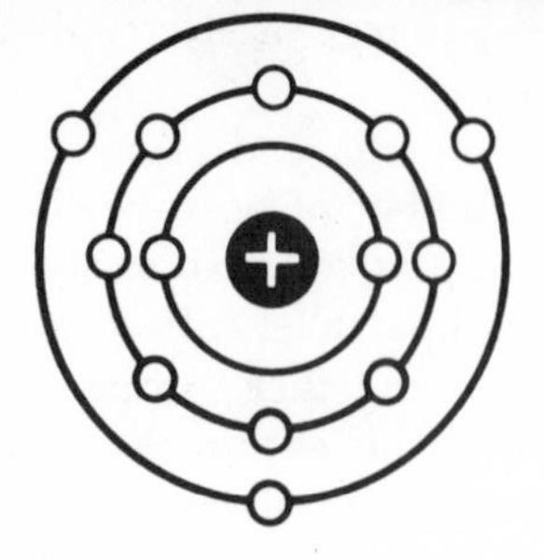

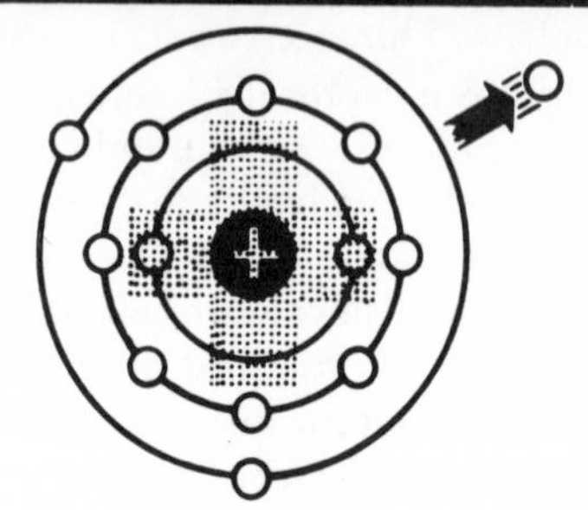

When an atom loses an electron, it loses a negative charge. The part of the atom left behind therefore ceases to be electrically balanced, for its nucleus remains as positive as before, but one of the balancing negative charges has gone. It is therefore left positively charged. This positively charged body is called a *positive ion.* In solid materials, the atoms are held in place by the crystalline structure of the material and therefore do not move as free electrons do. In liquids and gases, however, ions can move like electrons and contribute to the current flow.

You have learned that all matter is made up of electronic structures and that the motion of the electrons freed from the outer orbits of atoms is an electric current. Before you can go further in your study of electricity you will find out how the flow of electrons is confined to certain places by the use of different materials called *conductors* and *insulators* and about the nature of electric charges and magnetism. These are very important ideas that you will need for all of your studies in electricity, so it is important that they be learned as soon as is possible.

Review of Electricity—What It Is

Now stop and review what you have found out about electricity and the electron theory. Then you will be ready to learn about conductors, insulators, semiconductors, electric charges, etc.

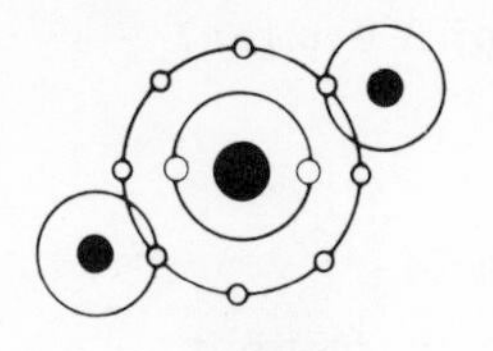

1. MOLECULE—The combination of two or more atoms. The smallest unit into which a substance—such as water—can be divided and still be identified as that substance.

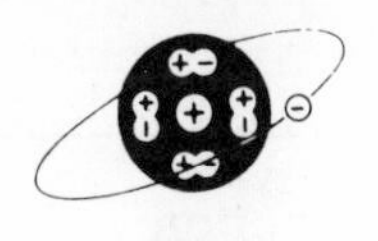

2. ATOM—The smallest particle into which an element—such as oxygen—can be divided and still retain its original properties.

3. NUCLEUS—The heavy positively charged central part of the atom.

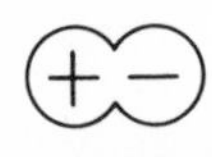

4. NEUTRON—The heavy neutral particles in the nucleus that behave like a combination of a proton and an electron.

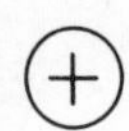

5. PROTON—The heavy positively charged particles in the nucleus.

6. ELECTRON—The very small negatively charged particles which are practically weightless and circle the nucleus in orbits.

7. BOUND ELECTRONS—Electrons in orbit in an atom.

8. FREE ELECTRONS—Electrons that have left their orbit in an atom and are wandering freely through a material.

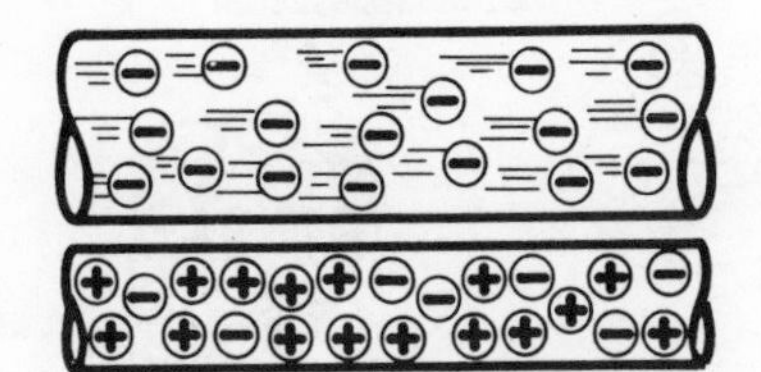

9. ELECTRIC CURRENT—The directional movement of free electrons.

10. POSITIVE CHARGE—A deficiency of electrons.

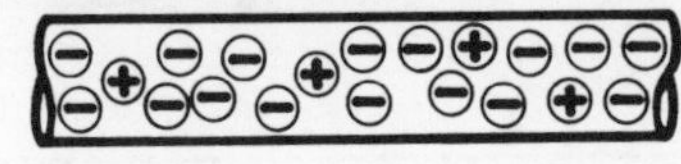

11. NEGATIVE CHARGE—A surplus of electrons.

Self-Test—Review Questions

1. What is the basic point of the electron theory?
2. Why is the electron theory still called a theory?
3. What is a molecule? An element?
4. Is the nucleus of an atom positively or negatively charged?
5. What is the charge on an electron? A proton? A neutron?
6. What are electrons?
7. What are free electrons?
8. Define positive and negative charges.
9. Define an electric current.
10. What is the difference between an electric charge and an electric current?

Learning Objectives—Next Section

Overview—You have learned that electricity is the flow of electrons. In the next section, you will learn about conductors, insulators, and semiconductors. The proper use of these makes current flow where we want it to.

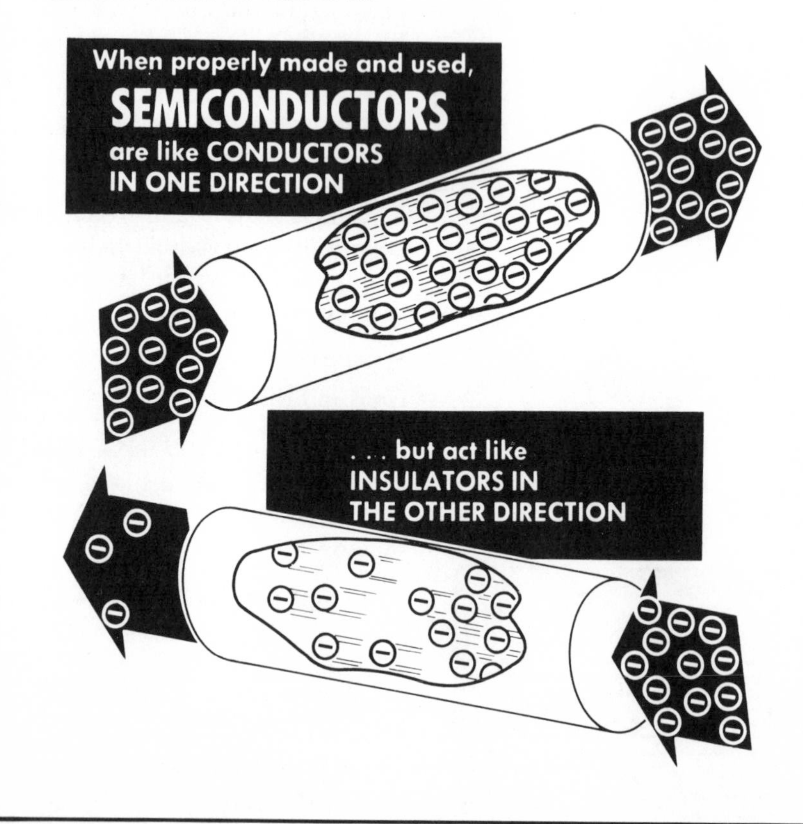

What a Conductor Is

You have learned that an electric current is the flow of electrons. Materials that permit the free motion of electrons are called *conductors*. Copper wire is considered a good conductor because it has many free electrons. The atoms of copper are held in place by the structure that copper forms when it is a solid. The electrons in the outer orbit of the copper atom are not very strongly bound and can be readily freed from the atom.

Electrical energy is transferred through conductors by means of the movement of free electrons that migrate from atom to atom inside the conductor. Each electron moves a very short distance to a neighboring atom where it replaces one or more of its electrons by forcing them out of its outer orbit. The replaced electrons repeat the process in other nearby atoms until the movement of electrons has been transmitted throughout the entire conductor. The more electrons that can be made to move in a material for a given applied force, the better conductor you have. Silver is the best conductor but we usually use copper, the next best, because it is cheaper. Recently, we have begun to use aluminum; when properly used it is almost as good a conductor as copper but has become much cheaper. Zinc, brass, and iron come next. In fact, most common metals are relatively good conductors. Salt water and similar solutions of salts or acids are also good conductors of electricity. Carbon is a good conductor, too.

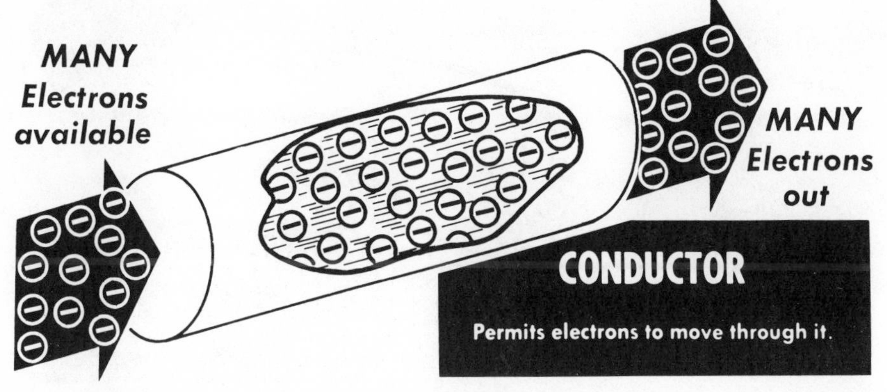

When you have learned more about electricity, you will learn how important it is to choose the right conductor of the right size to do a particular job. You will also learn that certain metals and alloys (mixtures of metals) are only fair conductors, although these materials are very useful, too.

When some metals are cooled to about —270 degrees Celsius (centigrade scale), they exhibit *superconductivity*. Under such conditions, these metals have essentially *no* resistance to the flow of electrons. Practical use is being made of superconductivity in cryogenic (supercold) electric motors and in strong electromagnets used in nuclear fusion work.

What an Insulator Is

Materials that have very few free electrons are *insulators*. In these materials, a lot of energy is needed to get the electrons out of the orbit of the atom. Even then only a few can be released at a time. Actually, there is no such thing as a perfect insulator. As a result, there is no sharp division between conductors and insulators; insulators can be thought of as poor conductors. Materials such as glass, mica, rubber, plastics, ceramics, and slate are considered to be among the best insulators. Dry air is also a good insulator. Another name for insulator is *dielectric*.

It may surprise you to know that insulators are just as important as conductors, because without them it would not be possible to keep electrons flowing in the places that we want and to keep electrons from flowing in places where we do not want them.

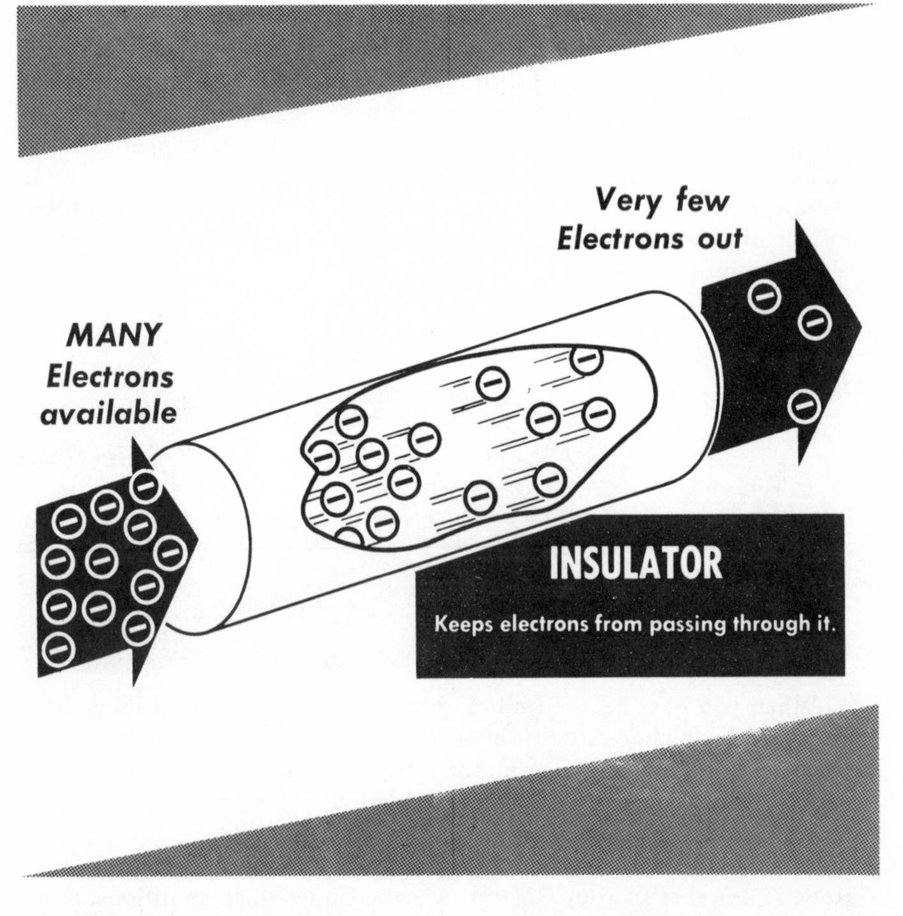

When you have learned more about electricity you will see how important it is to choose the right insulator of the right size to do a particular job.

What a Semiconductor Is

As the name implies, a *semiconductor* is a material that has some characteristics of both insulators and conductors. In recent years, these semiconductor materials have become extremely important as the basis for transistors, diodes, and other solid-state devices that you have probably heard about. Semiconductors are usually made from germanium or silicon, but selenium and copper oxide, as well as other materials, are also used. To make these materials into semiconductors, carefully controlled impurities are added to them during manufacture. The important thing about semiconductors is not that they are midway between insulators and conductors. It is that when properly made, they will conduct electricity in one direction better than they will in the other direction. As you will see later, this is an extremely valuable property that you can take advantage of in a number of ways. You do not know enough at the present time to be able to learn more about semiconductors. However, you will learn more about them when you have learned more facts about electricity.

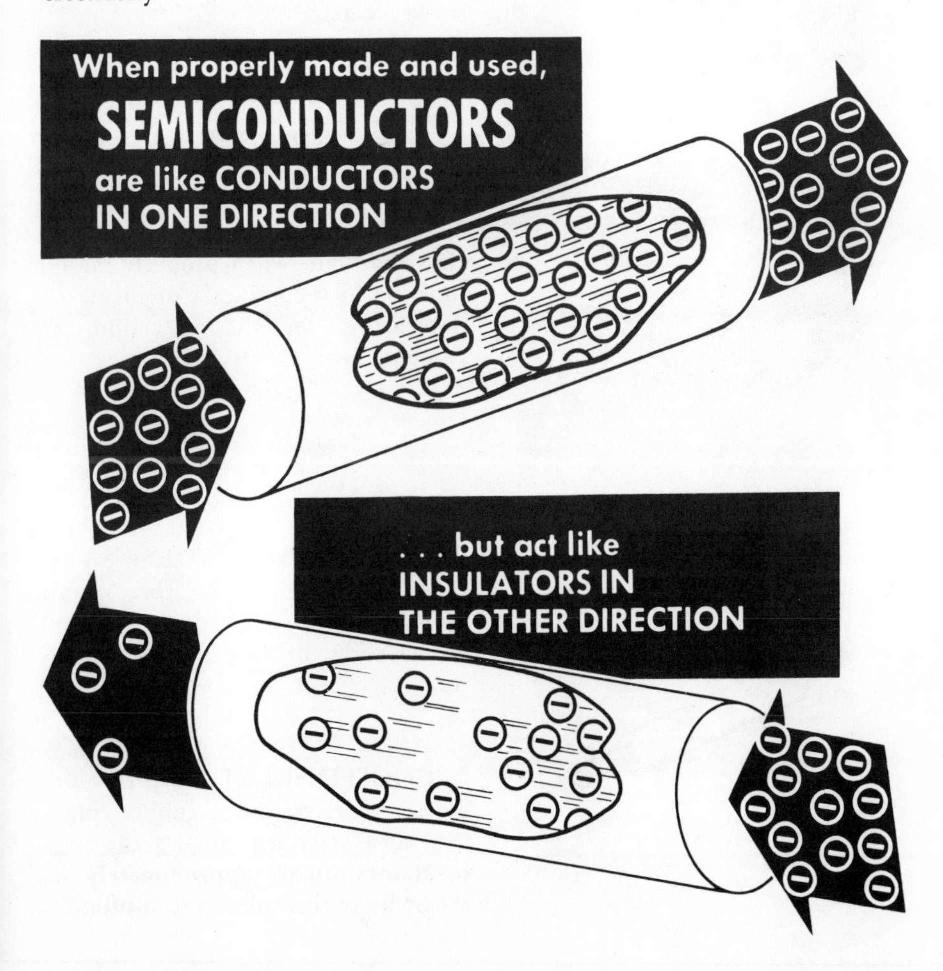

Review of Conductors, Insulators, Semiconductors

All materials can be classified as conductors, insulators, or semiconductors. There are no firm dividing lines. Also, there is no perfect conductor or perfect insulator. We use conductors and insulators in the right places to make electricity go where we want it and to keep it out of places where it should not be.

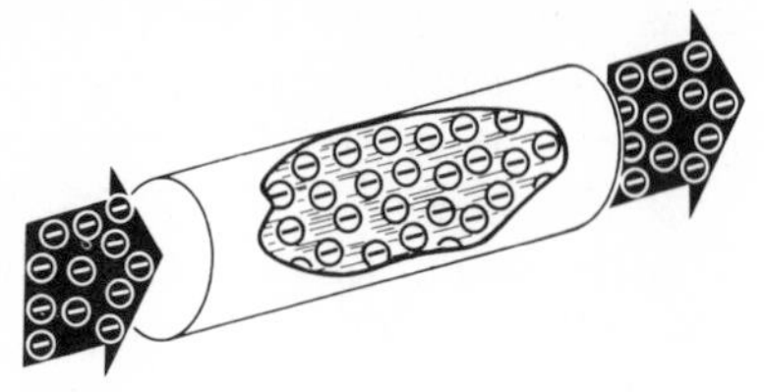

1. CONDUCTORS—Materials that permit the free movement of many electrons.

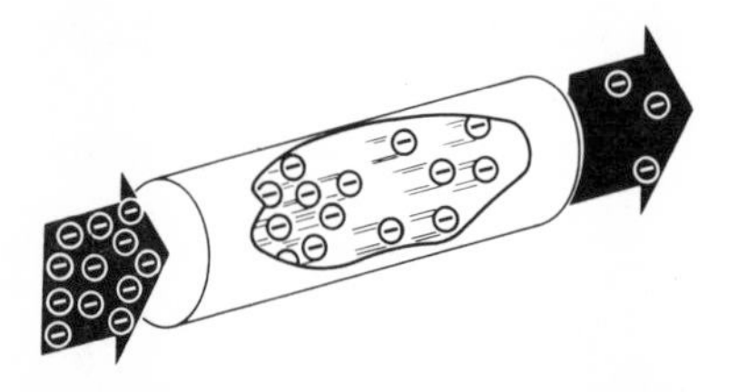

2. INSULATORS—Materials that do not permit the free movement of many electrons.

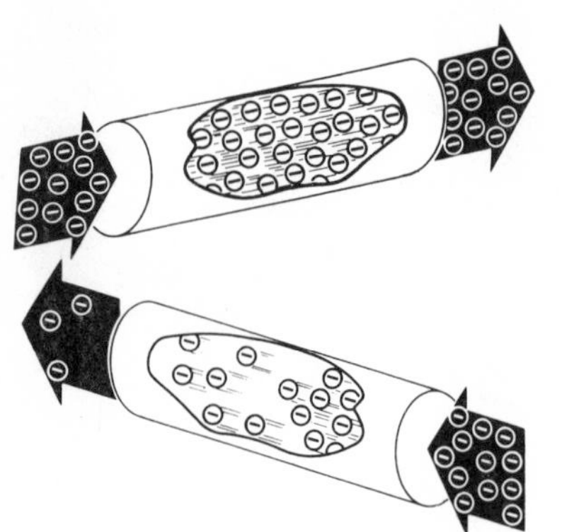

3. SEMICONDUCTORS — Materials that can, when properly made, function as a conductor or insulator depending on the direction of current flow.

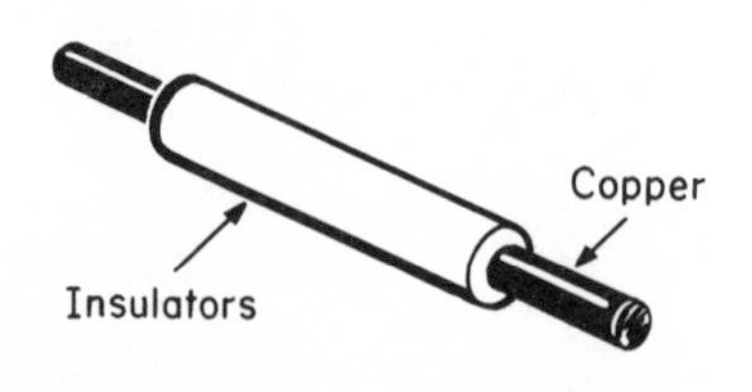

4. GOOD CONDUCTORS—Silver, copper, aluminum, zinc, brass, and iron are the best conductors, listed in the order of their ability to conduct.

5. GOOD INSULATORS—Dry air, glass, ceramics, mica, rubber, plastics, and slate are among the best insulators, listed approximately in the order of their ability to insulate.

Self-Test—Review Questions

1. What makes a good conductor?
2. Can materials other than metals be conductors?
3. Why is copper used as a conductor?
4. What makes a good insulator?
5. Describe some common insulators that you have seen.
6. Glass is a better insulator than rubber. Why then, do you find that rubber is very commonly used as an insulator?
7. What are the most important properties of semiconductors?
8. Compare semiconductors to conductors and insulators.
9. Are insulators as important in electricity as conductors? Why?
10. Choose a common electrical device with which you are familiar. Describe how the conductors and insulators are used. Why were the particular materials that were used chosen?

Learning Objectives—Next Section

Overview—Now that you know about conductors and insulators, you can learn about electric charges and static electricity. You will learn, in the next section how static charges can be generated and moved and how electric fields exist around a charged body.

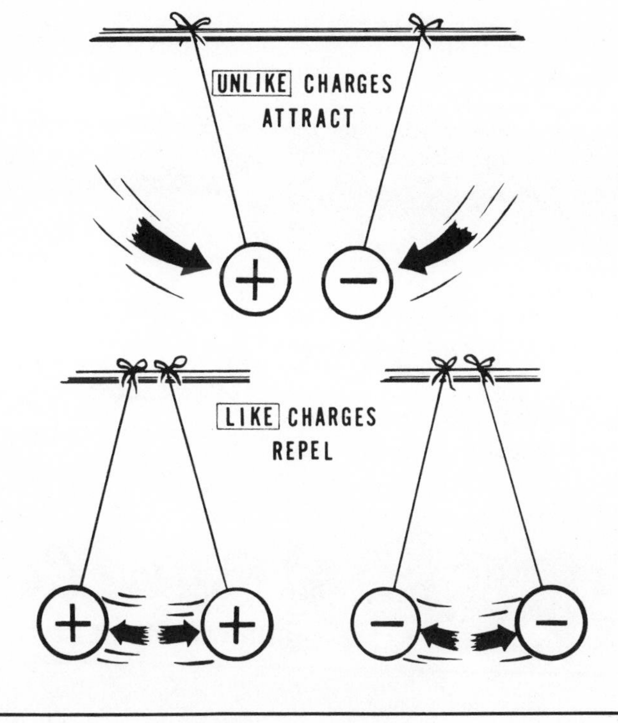

What Electric Charges Are

You have learned that electrons travel around the nucleus of an atom and are held in orbits by the attraction of the positive charge in the nucleus. When you force an electron out of its orbit, then the electron's action becomes what is known as electricity.

Electrons which are forced out of their orbits in some way will cause a lack of electrons in the material which they leave and will cause an excess of electrons at the point where they come to rest. This excess of electrons is called a *negative* charge and the lack of electrons is called a *positive* charge. When these charges exist and are not in motion, you have what is called *static electricity.*

To cause either a positive or negative charge, the electron must move while the positive charges in the nucleus do not move. Any material which has a positive charge will have its normal number of positive charges in the nucleus but will have electrons missing or lacking. However, a material which is negatively charged actually has an excess of electrons. Static electricity usually involves nonconductors since, if the materials were conductors, then the free electrons or negative charges could easily flow back toward the positive charges and the material would be neutral or uncharged.

You are now ready to find out how friction can produce this excess or lack of electrons to cause static electricity.

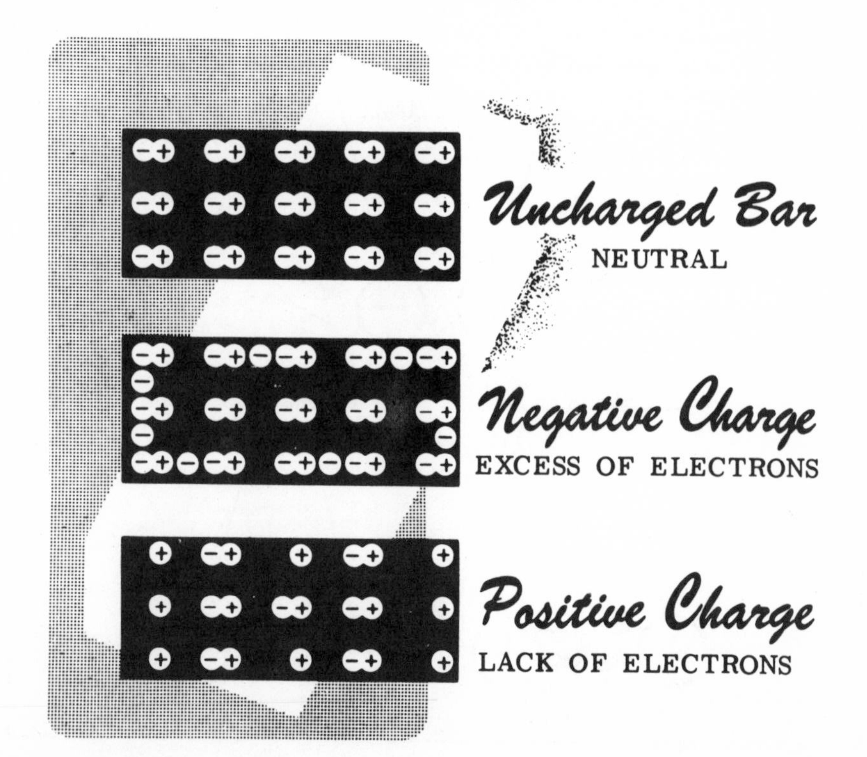

Static Charges from Friction

You have studied the electron and the meaning of positive and negative charges, so that you are now ready to find out how these charges are produced. The main source of static electricity is *friction*. If you rub two different materials together, electrons may be forced out of their orbits in one material and captured in the other. The material which captured electrons would then have a negative charge and the material which lost electrons would have a positive charge. If the materials are conductors, the electrons will move freely and the charges will be quickly neutralized. If the materials are insulators, however, then the charges will stay separated in the two materials.

When two materials are rubbed together, due to friction contact, some electron orbits at the surface of the materials will cross each other and one material may give up electrons to the other. If this happens, static charges are built up in the two materials, and friction has thus been a source of an electric charge. The charge could be either positive or negative depending on which material gives up electrons more freely.

Some materials which easily build up static electricity are glass, amber, hard rubber, wax, flannel, silk, rayon, and nylon. When hard rubber is rubbed with fur, the fur loses electrons to the rod—the rod becomes negatively charged and the fur positively charged. When glass is rubbed with silk, the glass rod loses electrons—the rod becomes positively charged and the silk, negatively charged. You will find out that a static charge may transfer from one material to another without friction, but the original source of these static charges is friction.

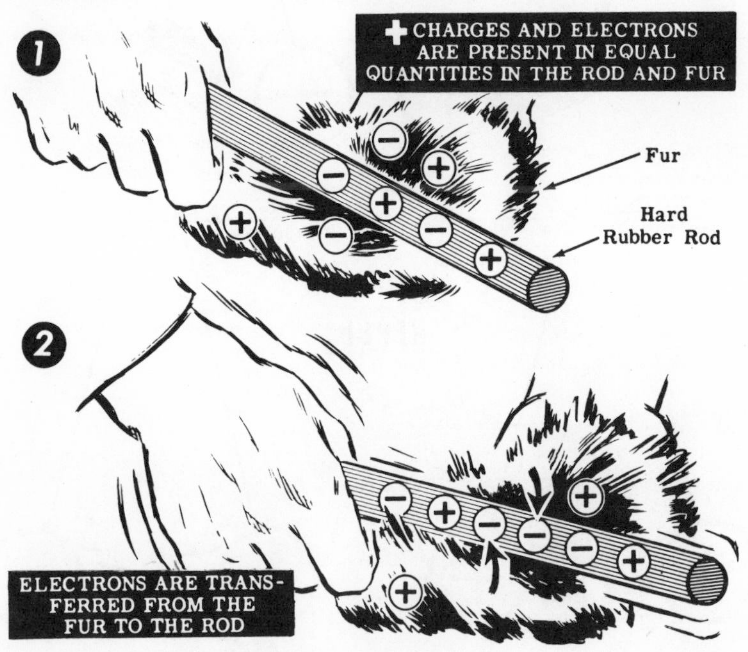

Attraction and Repulsion of Electric Charges

When materials are charged with static electricity, they behave in a different manner. For instance, if you place a positively charged ball near one which is negatively charged, the balls will attract each other. If the charges are great enough and the balls are light and free enough to move, they will come into contact. Whether they are free to move or not, a force of attraction always exists between unlike charges.

If you bring two materials of opposite charges together, the excess electrons of the negative charge will transfer to the material having a lack of electrons. This transfer or crossing over of electrons from a negative to a positive charge is called *discharge*, and by definition represents a current flow.

Using two balls with the same type of charge, either positive or negative, you will find that they repel each other.

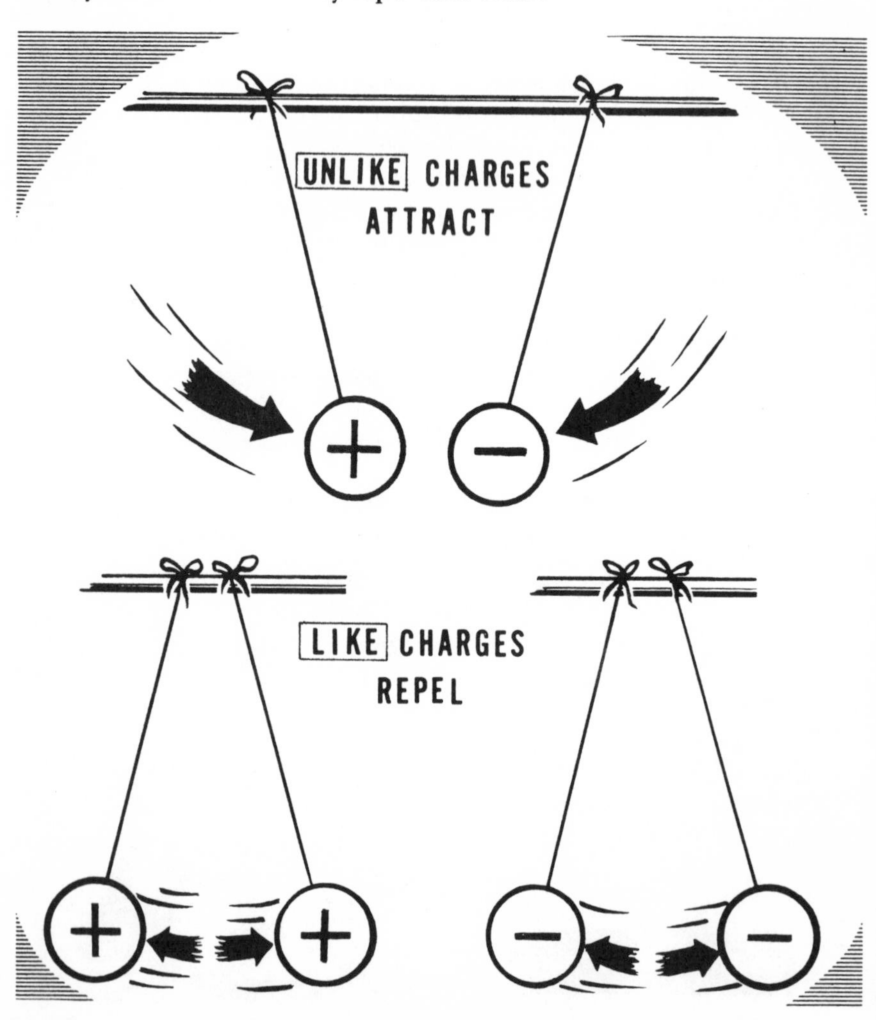

Electric Fields

You have learned that like charges repel each other and unlike charges attract. Since this happens when the charged bodies are separated, it must mean that there is a field of force that surrounds the charges and the effect of attraction or repulsion is due to this field. This field of force is called an *electric field of force*. It is also sometimes called an *electrostatic field* or a *dielectric field* since it can exist in air, glass, paper, a vacuum, or in any other dielectric or insulating material. Charles A. Coulomb, a French scientist, studied these fields in the 18th century and found that they behave in a predictable way according to what we now call *Coulomb's Law*. His law states that the force of attraction or repulsion between two charged bodies is proportional to the amount of charge present on both bodies divided by the square of the distance between them. Thus, the bigger we make the charge on our materials, the greater will be the attraction or repulsion between them; the further away we move the charged bodies, the less influence they will have on each other.

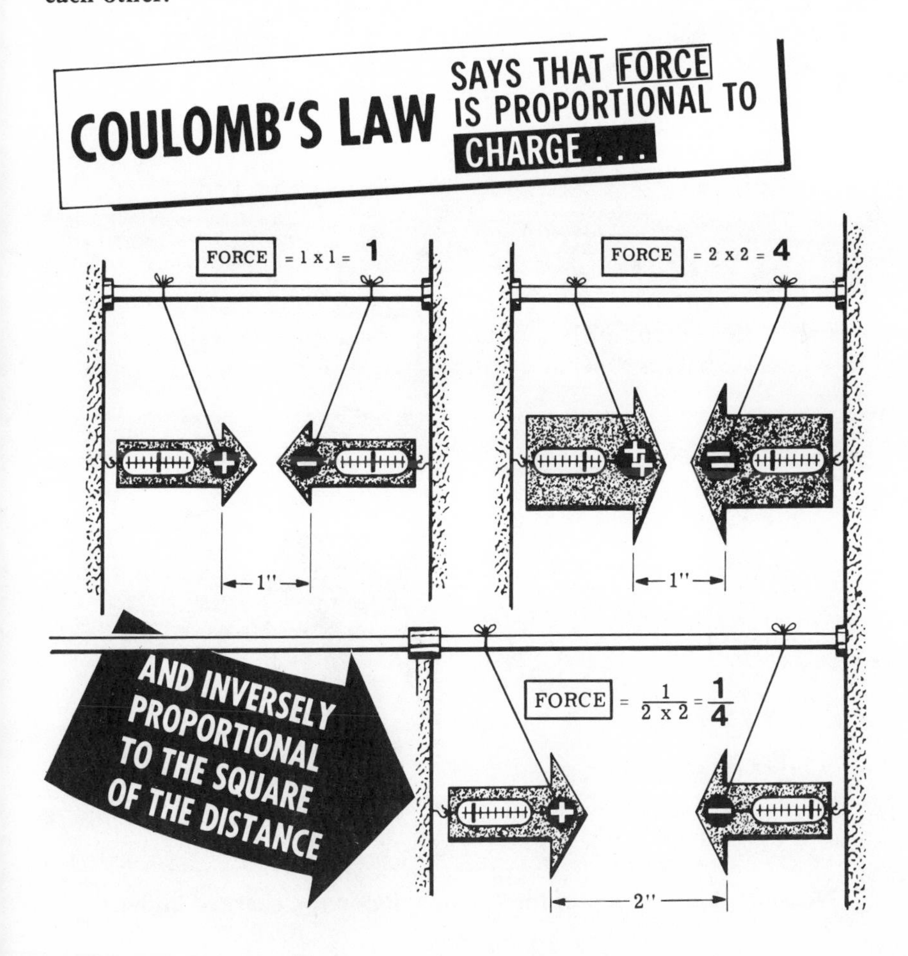

Electric Fields (continued)

The electric field around a charged body is usually represented by lines that are referred to as *electrostatic lines of force*. These lines are imaginary and are used to show the direction and strength of the field. Thus, they help us to understand what happens when these fields interact. To avoid confusion, the lines of force of a positive charge are always shown leaving the charge and the lines of force of a negative charge are always shown entering the charge.

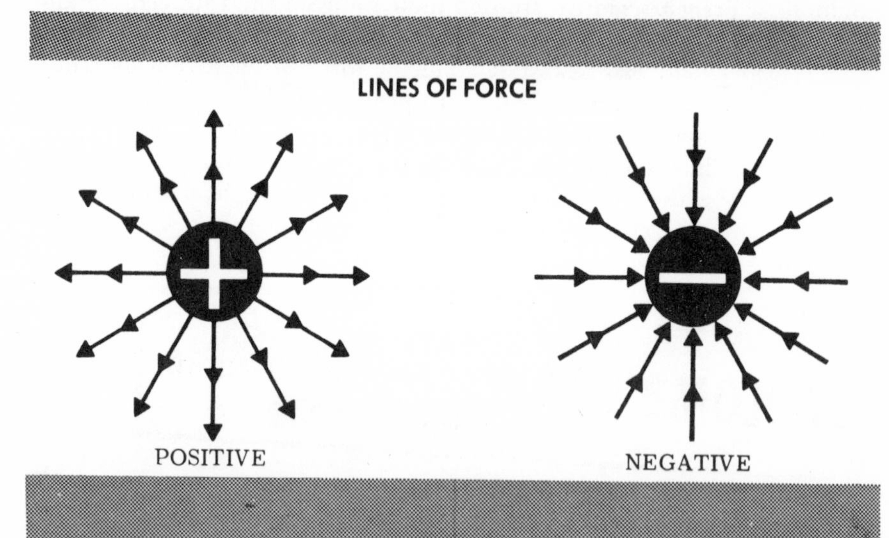

Using the concept of lines of force, you can now understand graphically why like charges repel and unlike charges attract.

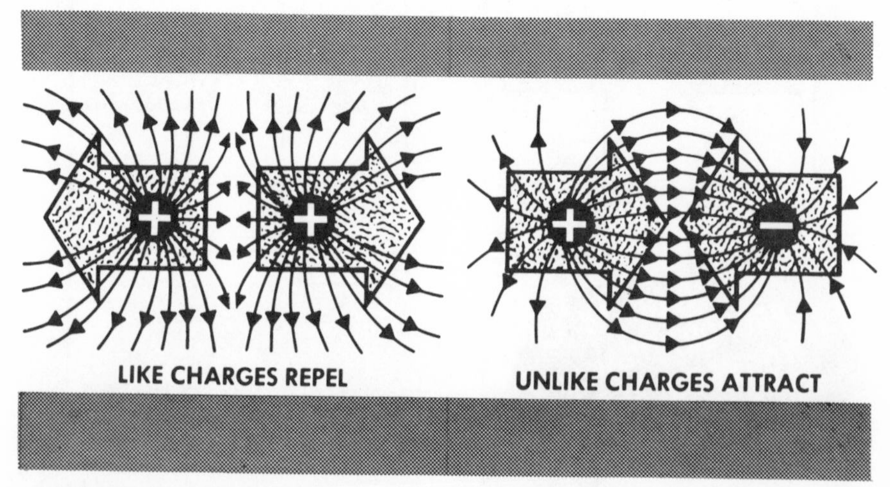

Note that all the lines of force terminate on the charged body.

Transfer of Electric Charges through Contact

Most electrostatic charges are due to friction. If an object has a static charge, it will influence other nearby objects. This influence may be exerted through contact or induction.

Positive charges mean a lack of electrons and always attract electrons, while negative charges mean an excess of electrons and always repel electrons.

If you should touch a positively charged rod to an uncharged metal bar that is supported on an insulator, it will attract electrons in the bar to the point of contact. Some of these electrons will leave the bar and enter the rod, causing the bar to become positively charged and decreasing the positive charge of the rod. When a charged object touches an uncharged object, it loses some of its charge.

In a similar way, the reverse happens when you start with a negatively charged bar.

POSITIVELY CHARGED ROD ALMOST TOUCHING UNCHARGED BAR

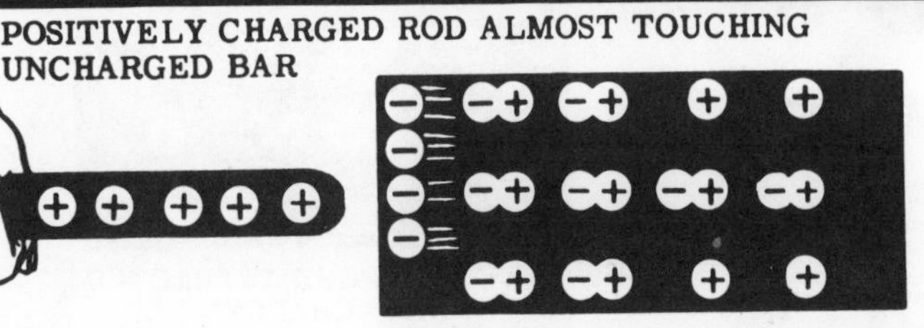

ELECTRONS ARE ATTRACTED BY POSITIVE CHARGE

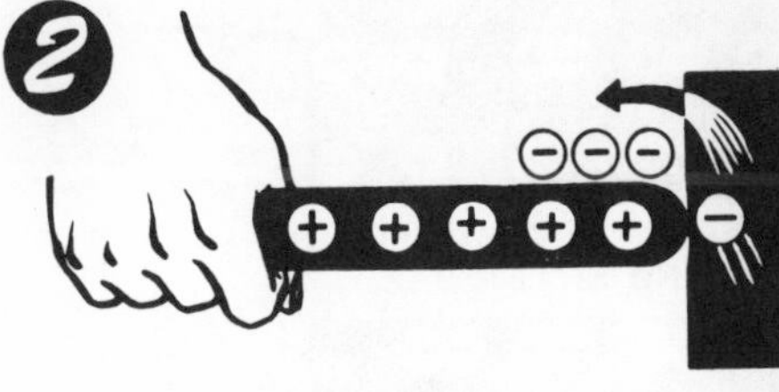

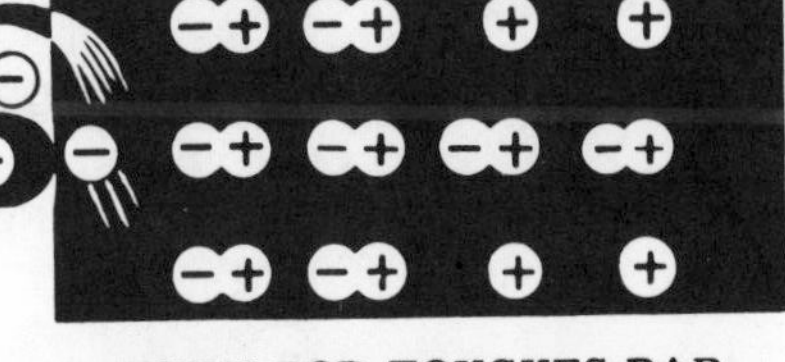

WHEN ROD TOUCHES BAR, ELECTRONS ENTER ROD

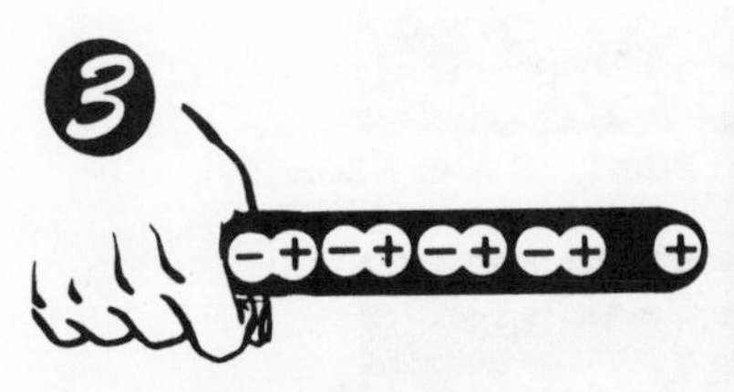

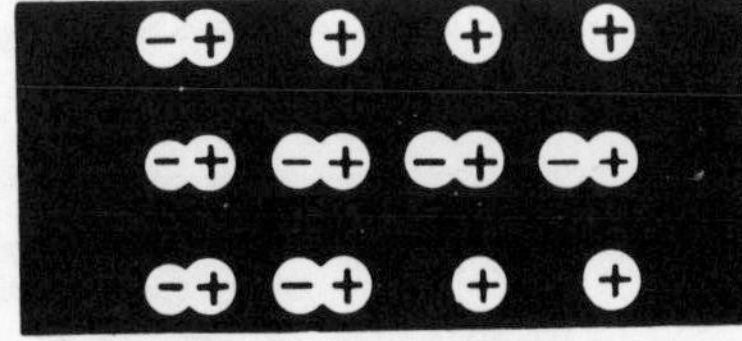

THE ROD IS NOW LESS POSITIVELY CHARGED

METAL BAR NOW HAS POSITIVE CHARGE

Transfer of Electric Charges through Induction

You have seen what happens when you touch a metal bar with a positively charged rod. Some of the charge on the rod is transferred and the bar becomes charged. Suppose that instead of touching the bar with the rod, you only bring the positively charged rod near to the bar. In that case, electrons in the bar would be attracted to the point nearest the rod, causing a negative charge to be induced at that point. The opposite side of the bar would again lack electrons and be positively charged. Three charges would then exist, the positive charge in the rod, the negative charge in the bar at the point nearest the rod, and a positive charge in the bar on the side opposite the rod. By allowing electrons from an outside source (your finger, for instance) to enter the positive end of the bar, you can give the bar a negative charge. This method of charge transfer is called *induction* because the charge distribution is induced by the presence of the charged rod rather than by actual contact.

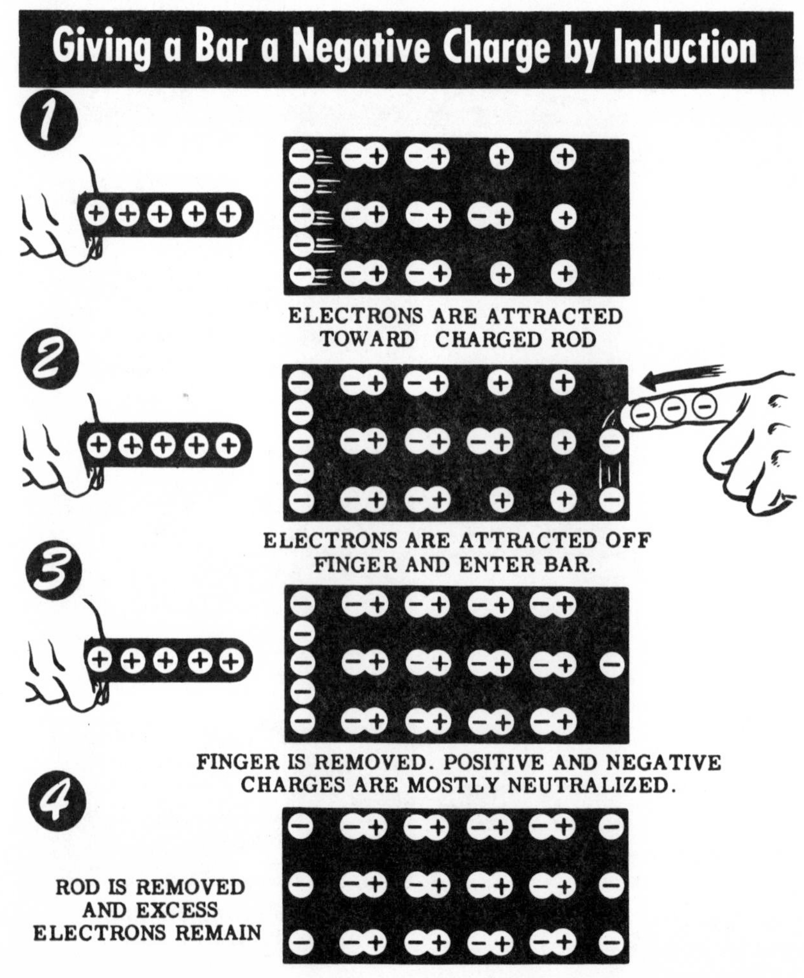

Discharge of Electric Charges

Whenever two materials are charged with opposite charges and placed near one another, the excess electrons on the negatively charged material will be pulled toward the positively charged material. By connecting a wire (conductor) from one material to the other, you would provide a path for the electrons of the negative charge to cross over to the positive charge, and the charges would thereby neutralize. Instead of connecting the materials with a wire, you might touch them together (contact), and again the charges would disappear.

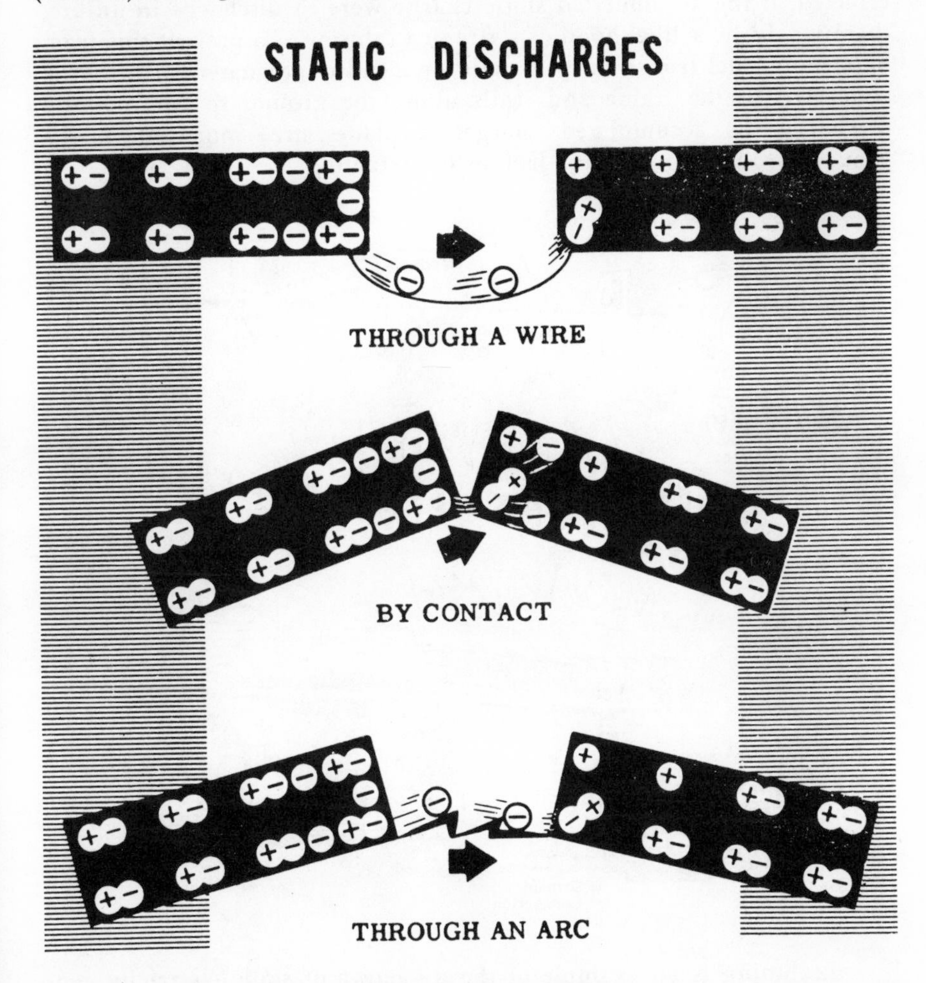

If you use materials with strong charges, the electrons may jump from the negative charge to the positive charge before the two materials are in contact. In that case, you would actually see the discharge in the form of an *arc*. With very strong charges, static electricity can discharge across large gaps, causing arcs many feet in length.

Discharge of Electric Charges (continued)

Although static electricity has limited practical use, its presence can be unpleasant and even dangerous if it discharges through an arc. You have probably had the experience of accumulating a static charge on a dry day and getting an unpleasant shock when you touched a metal object. Automobiles and trucks can pick up static charges from the friction of their tires on the road. Airplanes also can pick up static charges from the friction of their motion through the air. When a vehicle or truck is carrying an inflammable liquid such as gasoline, or an airplane is being refueled, if the accumulated static charge were to discharge in an arc, there would be a likelihood of a fire or explosion. To prevent this from happening, fuel trucks carry a chain or metal-impregnated strip that is connected to the frame and trails along the ground to continuously discharge the accumulated charge. Airplanes are connected to the ground through a grounding jack before refueling.

Lightning is an example of the discharge of static electricity generated from the friction between a cloud and the surrounding air. As you probably know, the energy in a stroke of lightning is enormous. Stationary objects, such as houses, can be protected from the effects of lightning by a lightning rod that minimizes the attracting (+) charge in the vicinity of the house.

Review of Electric Charges

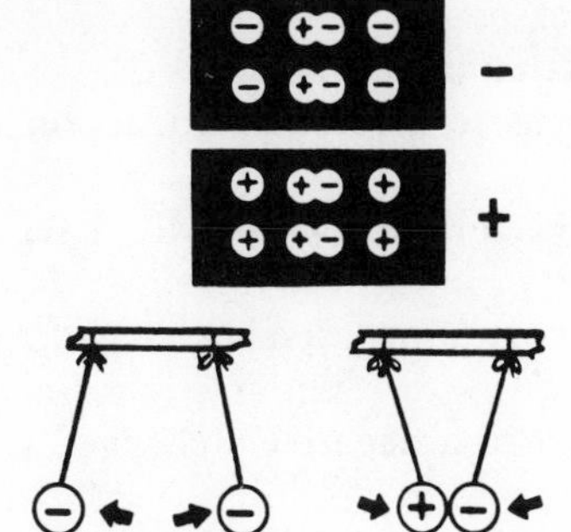

1. NEGATIVE CHARGE—An excess of electrons.

2. POSITIVE CHARGE—A lack of electrons.

3. REPULSION OF CHARGES—Like charges repel each other.

4. ATTRACTION OF CHARGES—Unlike charges attract each other.

5. STATIC ELECTRICITY—Electric charges at rest.

6. FRICTION CHARGE—A charge caused by rubbing one material against another.

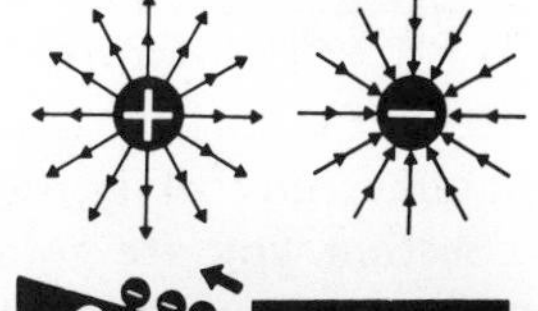

7. ELECTRIC FIELD—A field of force that surrounds a charged body.

8. CONTACT CHARGE—Transfer of a charge from one material to another by direct contact.

9. INDUCTION CHARGE—Transfer of a charge from one material to another without actual contact.

10. CONTACT DISCHARGE — Electrons crossing over from a negative charge to positive through contact.

11. ARC DISCHARGE—Electrons crossing over from a negative charge to positive through an arc.

12. COULOMB'S LAW—The force of attraction or repulsion is proportional to the amount of charge on each body and inversely proportional to the square of the distance between them.

Self-Test—Review Questions

1. Define a negative charge. A positive charge.
2. What are the rules for attraction and repulsion of charges?
3. According to Coulomb's Law, what happens to a force of attraction or repulsion when the distance is cut in half?
4. What concept do we use to account for the force between two charged bodies?
5. Describe, using diagrams, what happens when a negatively charged rod is used to charge a metal bar by contact.
6. Describe, using diagrams, what happens when a negatively charged rod is used to charge a metal bar by induction.
7. Describe two ways that a pair of charged bodies can be discharged.
8. Assume that you have two bodies, one with a negative charge and the other with a positive charge that is twice that of the negative body. What is the charge on each body after they have been discharged by each other?
9. What is the purpose of the grounding strap on gasoline trucks?
10. What is lightning?

Learning Objectives—Next Section

Overview—One of the most important effects of electricity is the generation of magnetic fields. To know about electricity, you must know about magnetism. In the next section, you are going to learn about the properties of magnets and how magnetic fields behave.

Natural Magnets

In ancient times, the Greeks discovered that a certain kind of rock, which they originally found near the city of Magnesia in Asia Minor, had the power to attract and pick up bits of iron. This rock was actually a type of iron ore called *magnetite*, and its power of attraction is called *magnetism*. Rocks containing ore that has this power of attraction are called *natural magnets*.

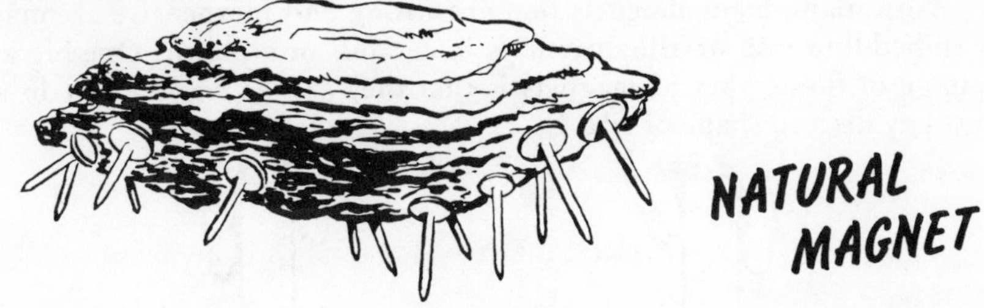

Natural magnets were seldom used until it was discovered that a magnet mounted so that it could turn freely would always turn so that one side would point to the north. Bits of magnetite suspended on a string were called *lodestones*, meaning a leading stone, and were used as crude compasses for desert travel by the Chinese more than 2,000 years ago. Crude mariner's compasses constructed of natural magnets were used by sailors in the early voyages of exploration.

The Earth itself is a large natural magnet, and the action of a natural magnet in turning toward the north is caused by the magnetism of the Earth.

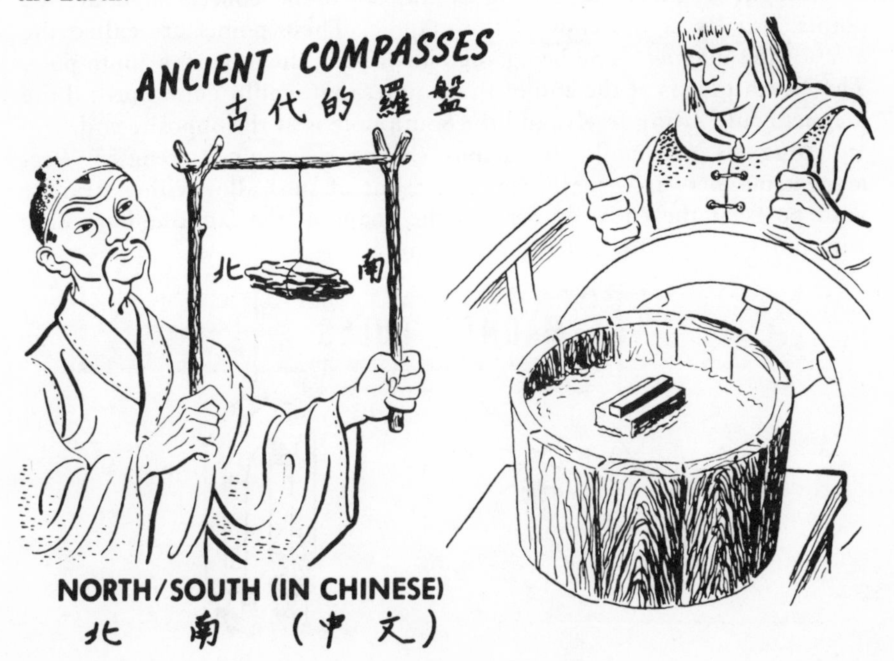

Permanent Magnets

In using natural magnets, it was found that a piece of iron stroked with a natural magnet became magnetized to form an *artificial magnet.* Artificial magnets may also be made electrically and materials other than iron may be used to form stronger magnets. Alloys containing nickel and cobalt make the best magnets and are usually used in strong magnets.

Nowadays, many magnets that are strong and inexpensive are made by embedding iron or alloy particles in ceramic or a plastic. One big advantage of these types of magnets is that they can be made easily in almost any desired shape or size.

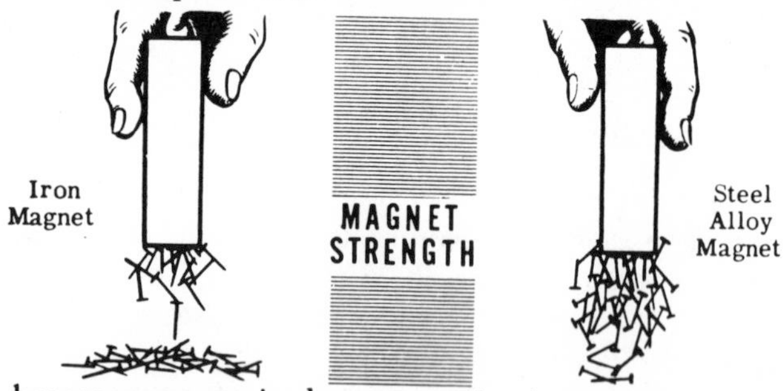

Iron becomes magnetized more easily than other materials, but it also loses its magnetism easily so that magnets of soft iron are called *temporary magnets.* Magnets made of steel alloys hold their magnetism for a long period of time and are called *permanent magnets.*

Magnetic effects in a magnet appear to be concentrated at two points, usually at the ends of the magnet. These points are called the *poles* of the magnet—one being the North pole, the other the South pole. The North pole is at the end of the magnet that would point north if the magnet could swing freely, and the South pole is at the opposite end.

Magnets are made in various shapes, sizes, and strengths. Permanent magnets are usually made of a bar of steel alloy, either straight with poles at the ends, or bent in the shape of the familiar horseshoe with poles on opposite sides of the opening.

The Nature of Magnetic Materials

Magnetism is a property shown by only a few types of materials, for example iron, cobalt, and nickel, and alloys containing these materials. Two questions that you might ask are (1) why only a few materials show magnetic properties, and (2) why these materials must be magnetized to become magnets? You can get an answer to these questions by looking at what happens when you take a bar magnet and break it into pieces.

If you did this, you would find that each of the pieces was a magnet, but of course, it would be much weaker. If you did the same thing with an unmagnetized bar of the same material, you would get small pieces of unmagnetized material. If, however, you could break the bar into very small pieces consisting of only a few million, million atoms, you would find that these very small pieces for both the magnetized and unmagnetized bar had magnetic properties.

Physicists tell us that the electrons that orbit the nucleus of an atom create a magnetic field in all atoms. In most materials, the electrons go in different directions and their fields cancel so the individual atoms have no net (resultant) magnetic field. Even most atoms with an odd number of electrons are nonmagnetic since these atoms are arranged in groups of about a million, million atoms called *domains*; and these atoms are arranged at random so there is no net (resultant) magnetic field (even though an individual atom with an odd number of electrons might be magnetic). In magnetic materials, the atoms do not all oppose each other in their orbit, in fact they add, so that each domain is strongly magnetic. In unmagnetized magnetic material, these domains are oriented randomly so that the magnetic fields from each are in all directions and there is no net (resultant) field. When, however, we stroke the material with another magnet or by other means, we align all the domains in one direction, the magnetic fields add together, and the bar becomes magnetic.

MAGNETIZED
Organized Orientation

UNMAGNETIZED
Random Orientation

We can see from this that a *permanent magnet* is one where the domains remain aligned when they have been aligned once; and a *temporary magnet* is one where the domains go back to the original random alignment when the aligning source has been removed.

Magnetic Fields

Magnetic fields and forces, just like electrostatic fields and forces, are invisible and can be observed only in terms of the effects they produce. In spite of this, the interaction of magnetic fields with each other and with conductors moving through them are among the most important things in electricity, since these effects are used to generate most of the electricity that we use and provide the power that can be derived from electricity.

The magnetic field about a magnet can best be explained as invisible lines of force leaving the magnet at one point and entering it at another. These invisible lines of force are referred to as *flux lines* and the shape of the area they occupy is called the *flux pattern*. The number of flux lines per unit area is called the *flux density*. When the flux density is measured in lines per square centimeter, we use a unit of flux density called the *gauss*, named for an 18th century physicist who investigated magnetism. The points at which most of the flux lines leave or enter the magnet are called the *poles*. The magnetic circuit is the path taken by the magnetic lines of force.

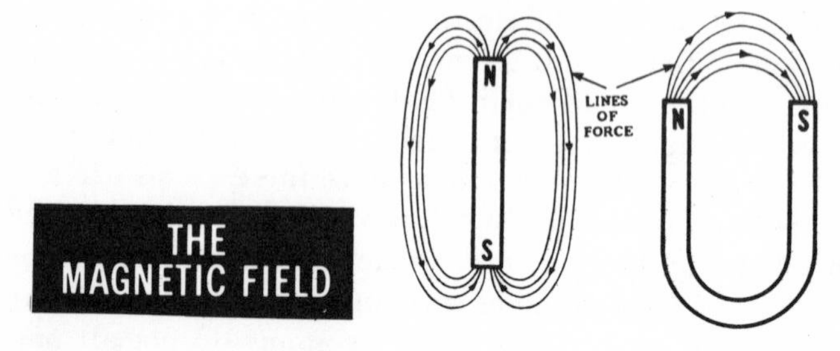

You can visualize the field around a magnet by using iron filings since the filings will become magnetized in the field of the magnet and then align themselves along the lines of force. If you place a sheet of paper or plastic over a magnet and then sprinkle the paper with iron filings, you will find that the filings will arrange themselves in a series of lines that do not cross and that terminate at the poles of the magnet. The concentration of filings will give an indication of the strength of the magnetic field at different points around the magnet.

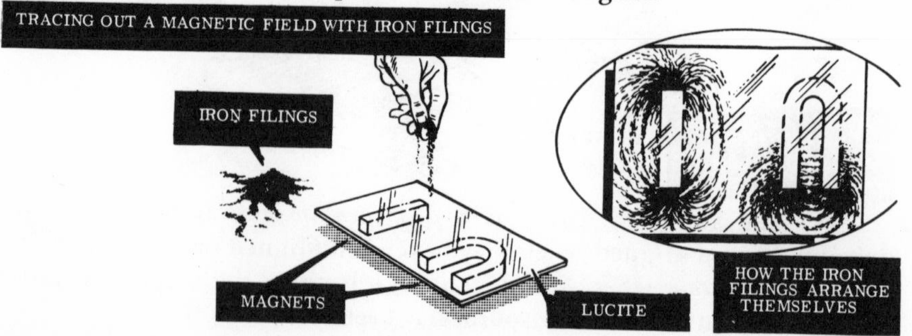

Magnetic Fields (continued)

If you were to bring two magnets together with the North poles facing each other, you would feel a force of repulsion between the poles. Bringing the South poles together would also result in repulsion, but if a North pole is brought near a South pole, a force of attraction exists. In this respect, magnetic poles are very much like static charges. Like charges or poles repel each other and unlike charges or poles attract. The laws of repulsion and attraction are like those for electric charges—that is, the force of repulsion or attraction is proportional to the strengths of the poles and inversely proportional to the distance between them.

The action of the magnetic poles in attracting and repelling each other is due to the magnetic field around the magnet. As has already been explained, the invisible magnetic field is represented by lines of force which leave a magnet at the North pole and enter it at the South pole. Inside the magnet the lines travel from the South pole to the North pole so that a line of force is continuous and unbroken.

One characteristic of magnetic lines of force is that they repel each other, never crossing or uniting. If two magnetic fields are placed near each other, as illustrated by the placement of the two magnets below, the magnetic fields will not combine but will reform in a distorted flux pattern. *Note that the flux lines do not cross each other.*

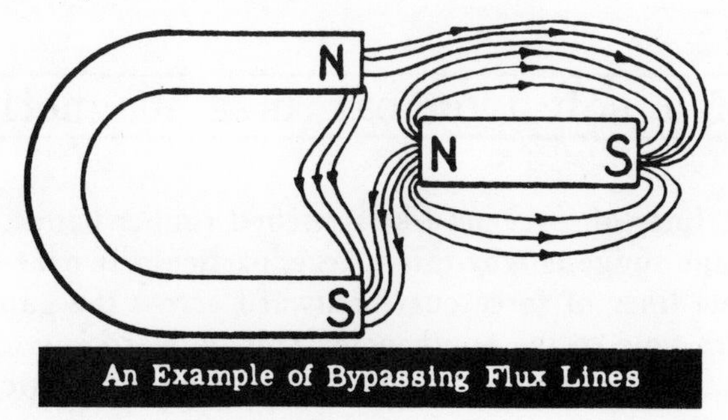

An Example of Bypassing Flux Lines

Magnetic Fields (continued)

There is no known insulator for magnetic lines of force. It has been found that flux lines will pass through all materials. Thus, most materials, except for magnetic materials, have no effect on magnetic fields. Conductors, insulators, air, or even a vacuum do not affect magnetic fields. However, they will go through some materials more easily than others. This fact makes it possible to concentrate flux lines where they are used, or to bypass them around an area or instrument.

On the previous page you were told that magnetic lines of force will go through some materials more easily than others. Those materials which will not pass flux lines so readily, or which seem to hinder the passage of the lines, are said to have a comparatively *high reluctance* to magnetic fields. Materials which pass or do not hinder the flow of flux lines are said to have a comparatively *low reluctance* to magnetic fields of force.

Magnetic lines of force take the path of least reluctance; for example, they travel more easily through iron than through air. Since air has a greater reluctance than iron, the concentration of the magnetic field becomes greater in the iron (as compared to air) because the reluctance is decreased. In other words, the addition of iron to a magnetic circuit concentrates the magnetic field which is in use.

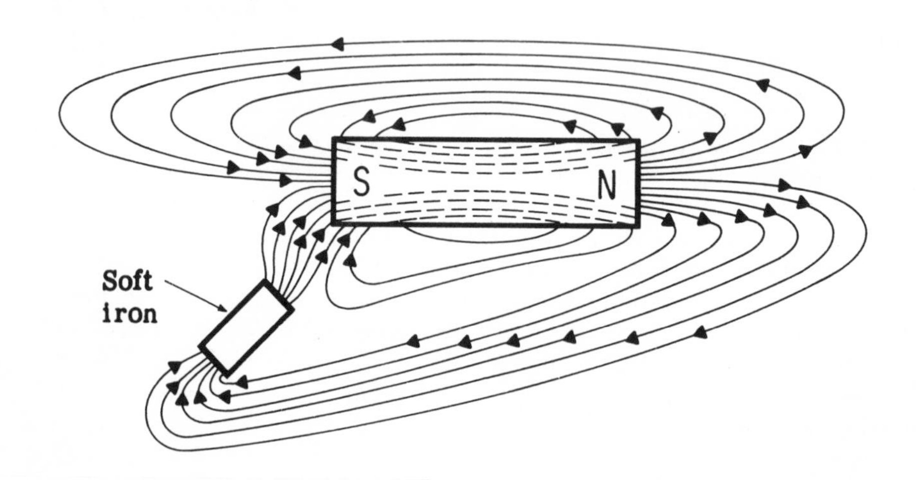

Effect of a soft iron bar in a magnetic field

Magnetic lines of force act like stretched rubber bands. The figure on the next page suggests why this is true, particularly near the air gap. Note that some lines of force curve outward across the gap in moving from the North pole to the South pole. This outward curve, or stretching effect, is caused by the repulsion of each magnetic line from

Magnetic Fields (continued)

its neighbor. However, the lines of force tend to resist the stretching effect and therefore resemble rubber bands under tension.

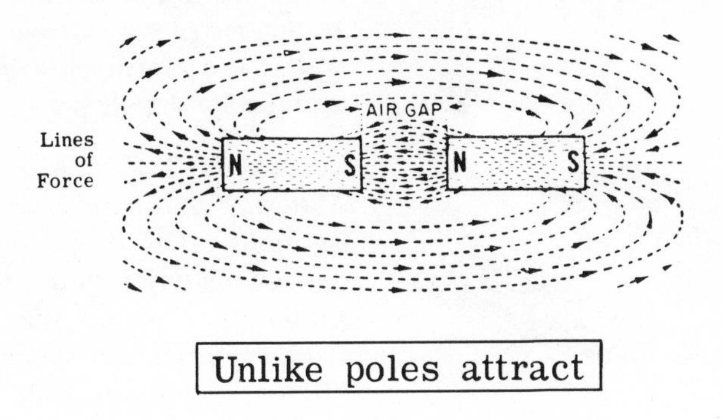

Unlike poles attract

As has already been mentioned, magnetic lines of force tend to repel each other. By tracing the flux pattern of the two magnets with like poles together in the diagram below, it can be seen why this characteristic exists.

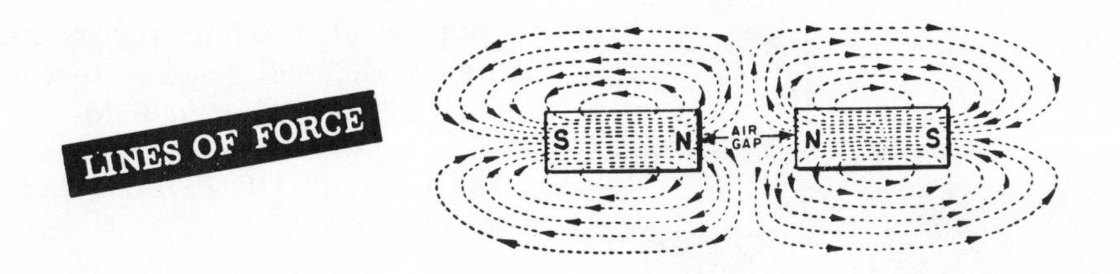

The reaction between the fields of the two magnets is caused by the fact that lines of force cannot cross each other. The lines, therefore, turn aside and travel in the same direction between the pole faces of the two magnets. Since lines of force which are directed in such a manner tend to push each other apart, the magnets mutually repel each other.

Only a certain number of magnetic lines can be crowded into a piece of material. This varies with each type of material. When the maximum number has been attained, the material is said to be *saturated*. This phenomenon is made use of in some pieces of electrical equipment but in most electrical equipment it is a very undesirable effect, since it limits the strength of a particular magnet made of a particular material. Thus, for equipment requiring very strong magnetic fields, a material with a very high magnetic saturation would be required, or it would be necessary to increase the amount of iron or other magnetic material.

Review of Magnetism

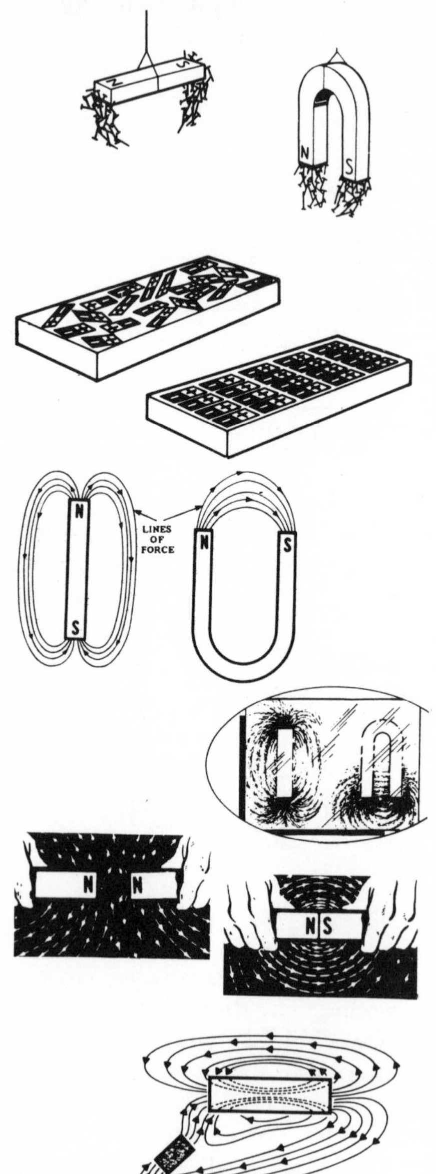

1. MAGNETIC POLES—Points on a magnet where there is a strong concentration of the magnetic field. If the magnet is allowed to swing freely, the North pole points north and the South pole points south.

2. MAGNETIC MATERIALS—Materials that have magnetic groups of atoms called domains. In unmagnetized material these domains are randomly oriented but in magnetized material the domains are all aligned in one direction.

3. MAGNETIC FIELD—Invisible lines of force that leave the magnet at the North pole and enter at the South pole. These lines are often called flux lines.

4. FLUX DENSITY—A measure of the number of flux lines per square centimeter that will give a picture of the strength of a magnetic field.

5. REPULSION OF POLES—Like poles repel.

6. ATTRACTION OF POLES—Unlike poles attract.

7. RELUCTANCE—A measure of the ease with which a material concentrates lines of force or flux lines. Materials with low reluctance tend strongly to concentrate flux lines. Magnetic lines of force take the path of least reluctance.

8. FLUX LINES DO NOT CROSS—Since flux lines do not cross because of the repulsion between them, they lie in parallel lines in a magnetic field.

Self-Test—Review Questions

1. What are permanent magnets? Temporary magnets?
2. Define a North Pole. A South Pole.
3. Based on what you know about the way magnetic materials work, do you think that it is necessary that all magnets have both a North and a South Pole? Why?
4. Draw a representation of the magnetic field around a bar magnet. A horseshoe magnet.
5. Of what use is the concept of flux density? Define it.
6. State the rules governing the interaction between magnetic fields or poles. Draw the fields for each case of like and unlike poles.
7. Why do you think saturation is a problem in some equipment that uses magnetism?
8. What is reluctance? Define high and low reluctance.
9. Can magnetic lines of force cross each other? Explain.
10. What happens to a magnetic field when another magnet is introduced? When a soft iron bar is introduced? Explain the effects in terms of what you have learned about magnetism.

Learning Objectives—Next Section

Overview—The study of electricity is mainly the study of current flow and the effects of current flow. You know that current flow is electrons in motion, now you will learn more about current flow.

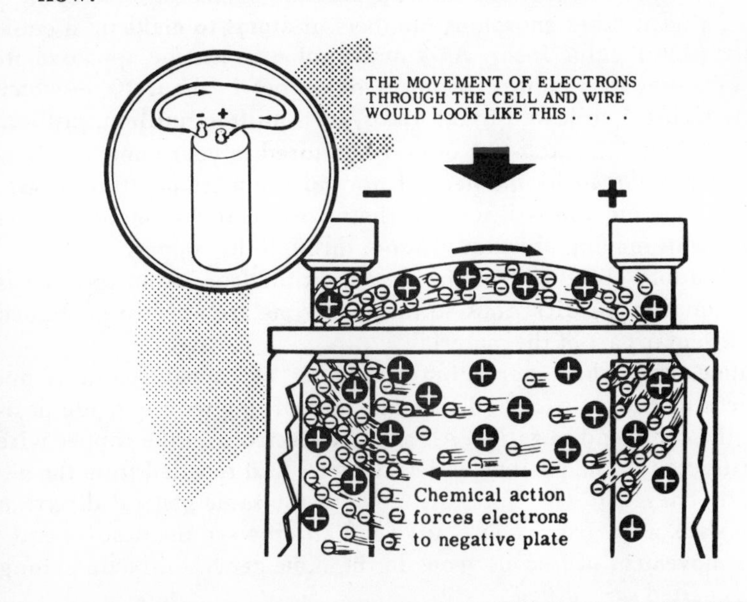

Electrons in Motion

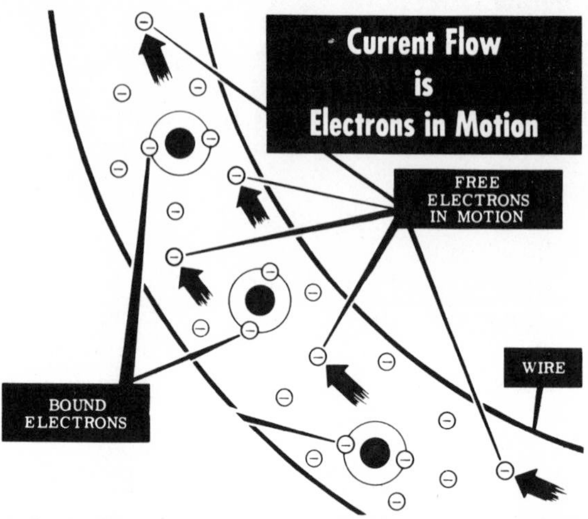

You already know that electrons in the outer orbits of an atom, being bound to the nucleus less tightly than electrons whose orbits are nearer the nucleus, can easily be forced from their orbits. You also know that in certain materials called *conductors* (they are generally, but not always, metals), very little energy is needed to liberate the outer electrons in this way.

In practice, the heat of normal room temperature is quite enough to liberate outer electrons in materials that are good conductors. The result is that a large number of electrons are normally *free* in these materials.

Now you must remember that an atom is something exceedingly small, and that it takes enormous numbers of atoms to make up a cubic centimeter (0.061 cubic inch). As a matter of interest, the approximate number of atoms in a cubic centimeter of copper is about 10^{24}—which means the figure 1 followed by 24 zeros, or a million, million, million, million! So if only one atom out of every hundred in your cubic centimeter of copper is forced by the heat of normal room temperature to give up a single electron, you will see that there will be an enormous number of free electrons moving about at random through the copper.

The random movement of the *free* electrons from atom to atom is normally equal in all directions so that electrons are not lost or gained by any particular part of the material.

Assume, now, that your cubic centimeter of copper is drawn out into a piece of copper wire and that one end of this wire is made positive, and the other end negative. All the free electrons in the copper wire will be attracted to the positive end of the wire and repelled from the negative end. They will, therefore, all move in the same general direction along the wire, away from the negative end and toward the positive end.

This movement of free electrons in the same general direction along the wire is called *current flow.*

Electrons in Motion (continued)

All electrons (being negative) are attracted by positive charges and are repelled by negative ones. They will always be attracted *from* a point having an excess of electrons, *toward* a point having a deficiency of them.

An electric cell or battery has exactly this property of having an excess of electrons at its negative terminal and a shortage of electrons at its positive terminal. This unbalance is maintained by chemical action as you will learn later.

Let's look at what happens when we connect a wire across the terminals of a cell. Instantly, a force will be exerted on all the free electrons in the wire, drawing some of them out of the end of the wire connected to the positive terminal of the cell. At the same time, the negative terminal of the cell will be pushing a lot more free electrons into the other end of the wire.

When electrons are drawn from one end of a piece of wire, there will result a *lack* of electrons (and therefore a positive charge) at that end. Similarly, when a lot of electrons are pushed by some outside source into the other end of the wire, there will be an *excess* of electrons (and, therefore, a negative charge) at that end. All these excess electrons will not only be repelled by one another, but also (and much more importantly) they will be attracted to the positive charge at the other end of the wire.

In this way, a continuous movement of electrons will take place *from the negatively charged* end of the wire *to the positively charged end*, for as long as electrons are furnished to one end of the wire and removed from it at the other end.

A battery (being a series of electric cells connected together) is a good way of maintaining current flow because, as you will learn, it is capable of furnishing a constant flow of electrons to its negative terminal, and of removing a constant stream of electrons from its positive terminal—and of continuing this for a long time.

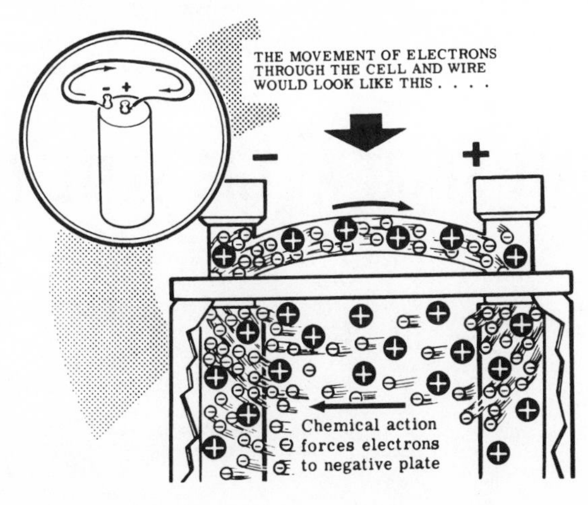

Electrons in Motion (continued)

By the way, do not think of electrons coming into direct physical contact with one another. You know that like charges repel, and that all electrons are negative. So, when a moving electron comes close to another electron, the second electron will be pushed away by the electric field of the first, without the two electrons themselves ever coming into contact.

When current flow starts in a wire, electrons start to move throughout the wire *at the same time*, just as the cars of a long train start and stop together.

If one car of a train moves, it causes all the cars of the train to move by the same amount, and free electrons in a wire act in the same manner. Free electrons are always present throughout the wire, and as each electron moves slightly it exerts a force on the next electron, causing it to move slightly and, in turn, to exert a force on the next electron. This effect continues throughout the wire.

When electrons move away from one end of a wire it becomes positively charged, causing all the free electrons in the wire to move in that direction. This movement, taking place throughout the wire simultaneously, moves electrons away from the other end of the wire and allows more electrons to enter the wire at that point.

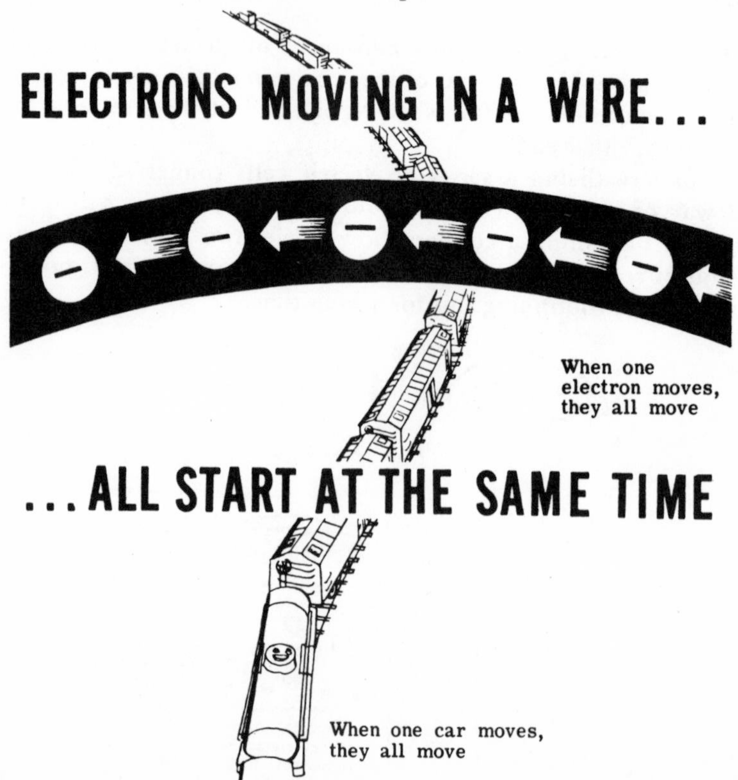

Direction of Current Flow

According to the electron theory, current flow is always from a negative (—) charge to a positive (+) charge. Thus, if a wire is connected between the terminals of a battery, current will flow from the (–) terminal to the (+) terminal.

Before the electron theory of matter had been worked out, electricity was in use to operate lights, motors, etc. Electricity had been harnessed but no one knew exactly why it worked. It was believed that an electric fluid moved in the wire from (+) to (—). This conception of current flow is called *conventional current flow.* You will find the conventional flow of (+) to (—) is often used in working with electrical and electronic equipment. Actually, it doesn't matter which direction you choose as long as you are consistent in working out any particular problem.

In all of our studies in electricity, we will be consistent and always use the direction of electron flow as the direction of current flow: that is, current flow is from *negative* to *positive.*

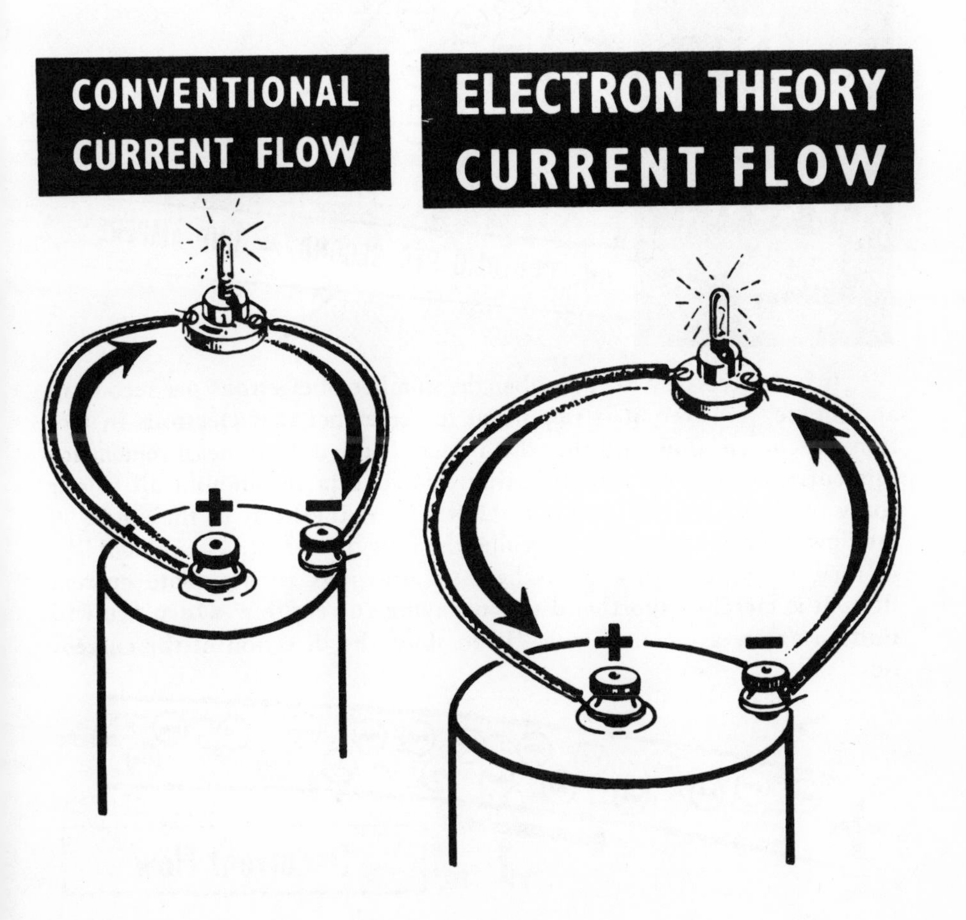

Units of Current Flow

Current flow, as you have learned, is the movement of electrons through a material. We measure the flow of current by measuring the number of electrons that flow past a given point in a given period of time. Since the coulomb is a measure of the number of electrons present, we can use it as the basis for the measurement of current flow. A coulomb is defined as about six and a quarter million, million, million electrons (or exactly 6.289×10^{18} electrons in mathematical terms). The unit of current flow is the ampere, which is defined as 1 coulomb flowing in 1 second. Thus, 1 ampere is a current flow of 1 coulomb per second and 2 amperes is a current flow of 2 coulombs per second, etc.

It isn't necessary to remember the number of electrons per second in an ampere; however, it is important to remember that electrons in motion are current flow and that the ampere is the unit of measurement for this current flow. We will be using this concept throughout all of our study of electricity. The study of electricity is the study of the effects of the flow of current and the control of the flow of current. The symbol "I" is used in calculations and schematic drawings to designate current flow. It is merely a shorthand way of saying current flow. Often you will find an arrow associated with "I" to show the direction of the current flow.

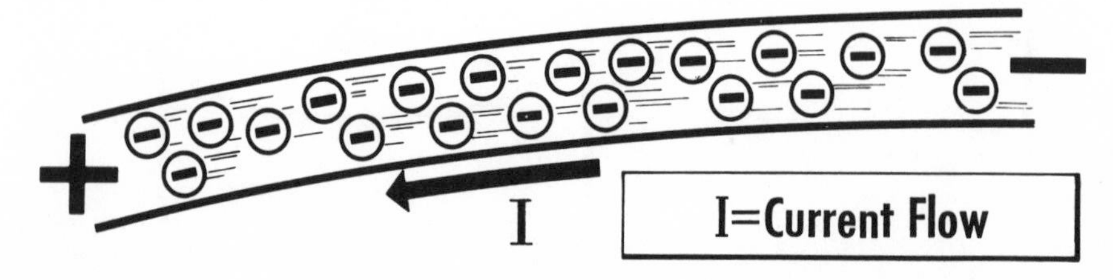

Review of Current Flow

Current flow does all the work involved in the operation of electrical equipment, whether it be a simple light bulb or some complicated electronic equipment such as a radio receiver or transmitter. For current to flow, a continuous path must be provided between the two terminals of a source of electric charges. Now suppose you review what you have found out about current flow.

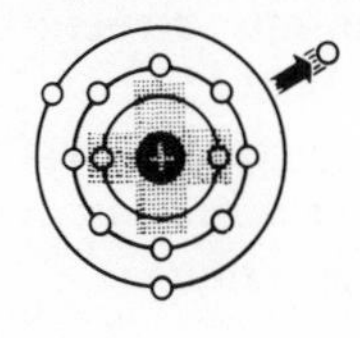

1. FREE ELECTRONS—Electrons in the outer orbits of an atom which can easily be forced out of their orbits.

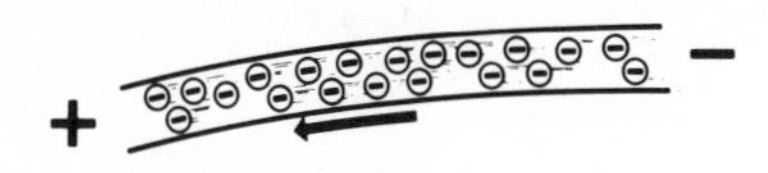

2. CURRENT FLOW—Movement of *free* electrons in the same direction in a material.

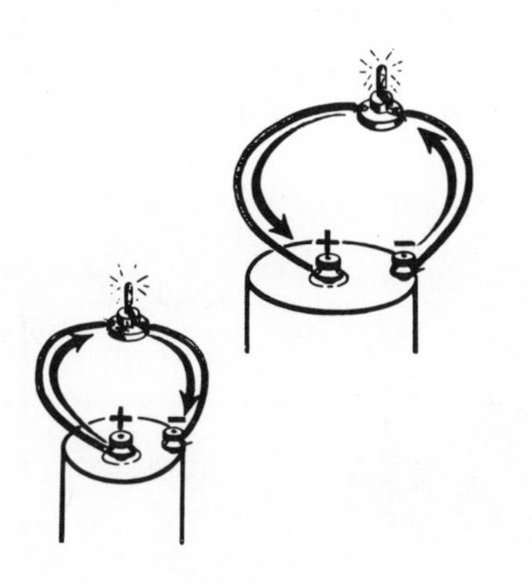

3. ELECTRON CURRENT—Current flow from a *negative* charge to a *positive* charge.

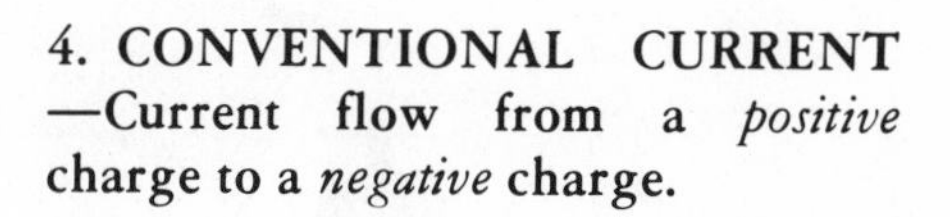

4. CONVENTIONAL CURRENT—Current flow from a *positive* charge to a *negative* charge.

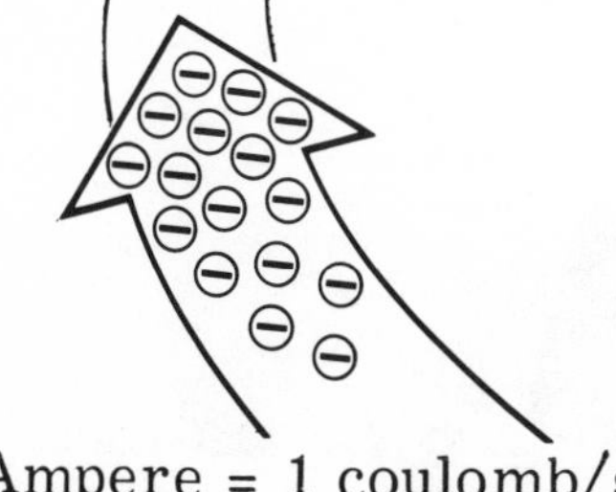

5. AMPERE—The unit of measure of current flow. It is equal to 1 coulomb per second.

I=Current

6."I"–The symbol used to designate current in schematic drawings and formulas.

Self-Test—Review Questions

1. What are free electrons?
2. Conductors have many free electrons. Why is it that they don't have an electric charge?
3. What is current flow?
4. Describe what happens when a piece of wire is connected across the terminals of a battery or a cell.
5. Describe current flow in a wire in terms of the atomic structure.
6. Do all of the electrons in a wire move together? Why?
7. What is the difference between electron theory current flow and conventional current flow? Does it make any difference which is used? Which will we use throughout the rest of our study of electricity?
8. What is the unit of measure of current flow?
9. How is it defined?
10. What is the symbol for current flow?

Learning Objectives—Next Section

Overview—Now that you know about current flow, you will learn about the *force* that makes current flow. This force called *electromotive force (or EMF)* must be maintained if current flow is to be maintained and it takes work (or energy) to do this, as you will see.

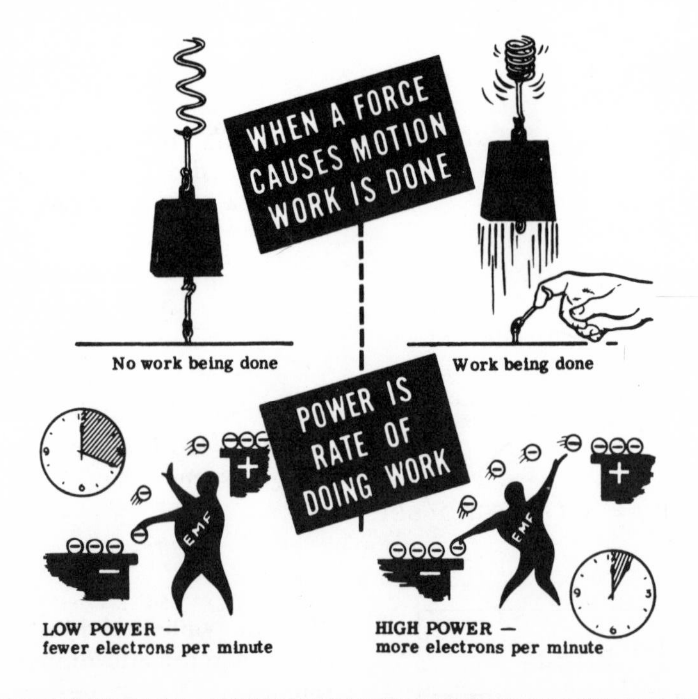

What Work Is

You have learned that the flow of electrons is an electric current and that its unit of measure is the ampere, which is equal to a flow of 1 coulomb per second. Energy is required to obtain or maintain a *charge difference* between two points. When a conductor is connected between these two points, a current will flow. The expenditure of energy is called *work* and like most things has a unit of measure. In the English system of measure based on pounds and feet, the unit of work is the *foot pound*; that is, the energy required to raise 1 pound a distance of 1 foot. On the other hand, if we had a 1-pound weight that already was raised off the ground 1 foot, we could connect it to a mechanism and get that amount of work back if we let it fall the 1 foot. When the weight is poised to do work, we say that it has *potential energy*, that is, the potential to do work. The meter and the gram are the units of distance and weight used by scientists in the metric system, which is being adopted worldwide. In the metric system the unit of work has been given a special name—the *joule*. A joule is equal to about ¾ of a foot pound (to put it exactly, 1 joule = 0.7376 foot pound).

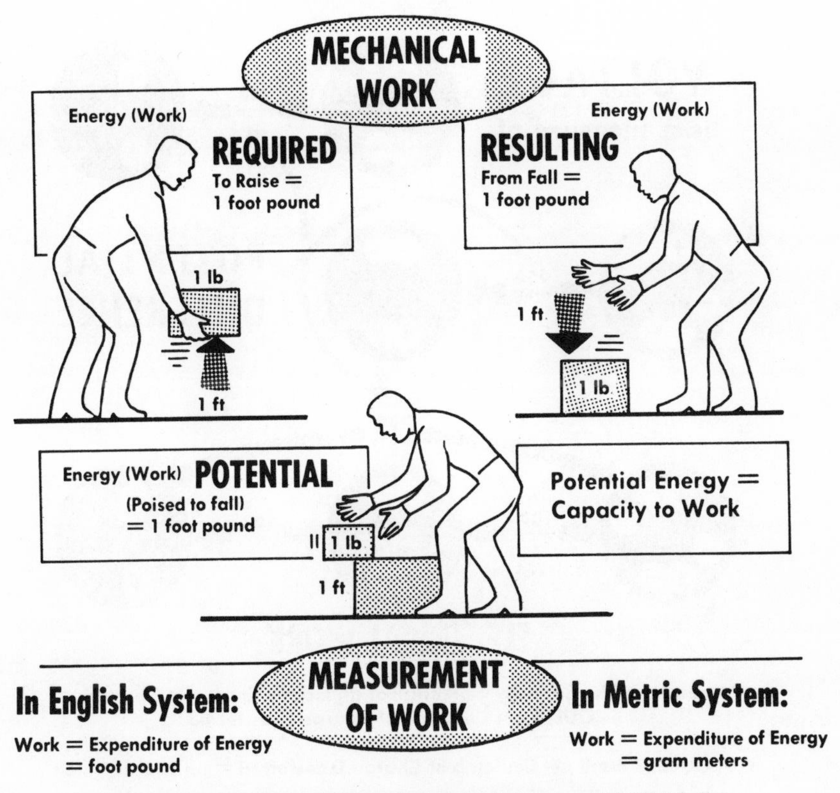

What EMF (Electromotive Force) Is

You may ask what all this discussion about work has to do with electricity, so let's see what the connection is. Since it is known that work is required to separate electric charges, then it follows that work results when these charges come back together. The separated charges, or potential difference, represent the *potential* (capacity) to do work just like our weight poised off the ground. As you shall see in the next section, several kinds of energy can be used to generate electricity; that is, to maintain potential difference. It is this potential difference that makes electrons move and thus do work or be used to generate other forms of energy.

The unit of potential difference is the *volt*. A volt is defined as the potential difference necessary to obtain 1 joule of work when one coulomb of charge flows. Thus, we have an *electromotive force* (EMF)—electron moving force—or potential difference of one volt when this happens.

We measure potential difference or emf in volts and we call the measured difference *voltage*. (And because emf is measured in volts, emf is often called, or is equated with, *voltage*.) The symbol that we use for voltage is "E" or "V."

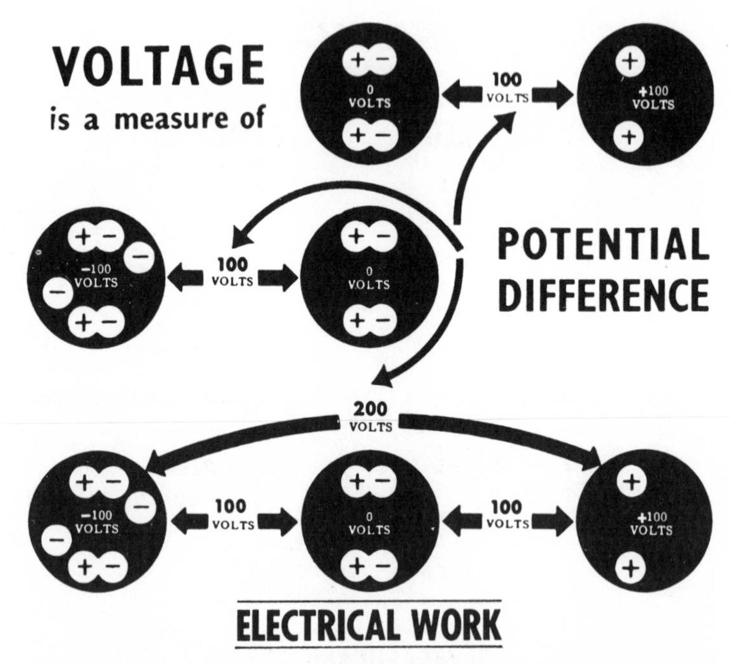

Measure For The Capacity (potential) of an Electric Charge To Do Work = JOULES per Coulomb of Charge Transferred

1 JOULE of Work per Coulomb of Charge Transferred = an EMF of 1 *Volt*

VOLTAGE = MEASURED POTENTIAL DIFFERENCE = *E* or *V*

What Electric Power Is

As you have just learned, whenever a force of any kind causes motion, *work* is done. When a mechanical force, for instance, is used to lift a weight, work is done. A force which is exerted *without* causing motion, however, such as the force of a spring held under tension between two objects which do not move, does *not* cause work to be done.

You know that a difference in potential between any two points in an electric circuit gives rise to a *voltage* which (when the two points are connected) causes electrons to move and current to flow. Here is an obvious case of a *force* causing motion, and causing *work* to be done. *Whenever a voltage causes electrons to move, therefore, work is done in moving them.*

The *rate* at which the work is done when moving electrons from point to point is called *electric power.* (It is represented by the symbol "P.")

The basic unit in which electric power is measured is the *watt.* It can be defined as the rate at which work is being done in a circuit in which a current of 1 ampere is flowing when the emf applied is 1 volt.

How EMF Is Maintained

In order to cause continuous current flow, electric charges must be maintained so that difference of potential, or voltage, exists all all times. At the terminals of a battery, for example, this difference is caused by the chemical action within the battery, and as electrons flow from the (—) terminal to the (+) terminal, the chemical action maintains this difference. The electric generator in your car acts in the same manner, with the action of a wire moving through a magnetic field maintaining a charge difference at the generator terminals. The energy to move the wire through the magnetic field of the generator comes from the engine. The voltage difference across the generator or battery terminals remains constant, and the charges on the terminals never become equal as long as the chemical action continues in the battery, and as long as the generator wire continues to move through the magnetic field.

A BATTERY MAINTAINS EMF

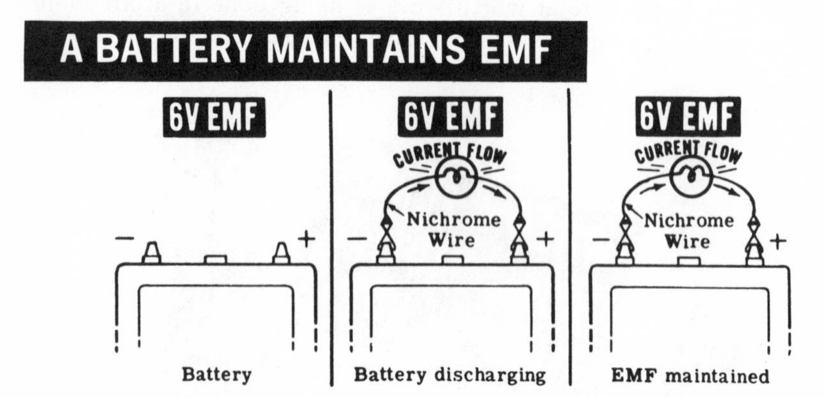

If the charge difference were not maintained at the terminals, as in the case of the two charged bars shown below, current flow would cause the two charges to become equal as the excess electrons of the (—) bar moved to the (+) bar. The voltage between the terminals would then fall to 0 volts, and current flow would no longer take place.

CHARGED BARS CANNOT

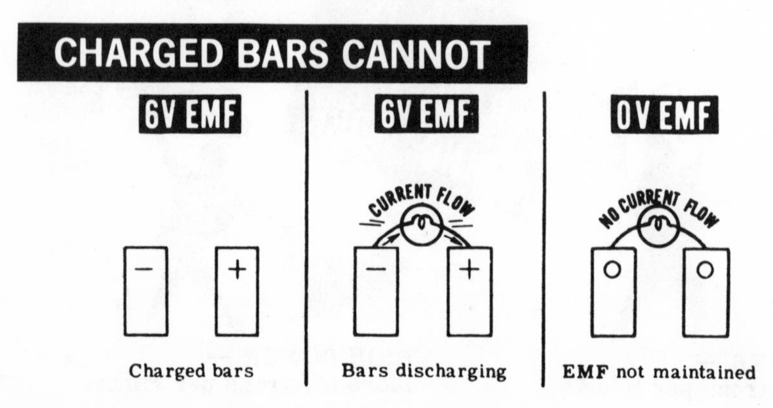

Voltage and Current Flow

Whenever two points of unequal potential or voltage are connected, current flows. The greater the emf or voltage, the greater the amount of current flow. Electrical equipment is designed to operate with a certain amount of current flow, and when this amount is exceeded, the equipment may be damaged. You have seen all kinds of equipment such as electric lamps, motors, radios, etc., with the voltage rating indicated. The voltage will differ on certain types of equipment, but it is usually 120 volts here in North America. This rating on a lamp, for example, means that 120 volts will cause the correct current flow. Using a higher voltage will result in a greater current flow and *burn out* the lamp, while a lower voltage will not cause enough current flow to make the lamp light up normally. While current flow makes equipment work, it takes emf or voltage to cause the current to flow, and the value of the voltage determines how much current will flow.

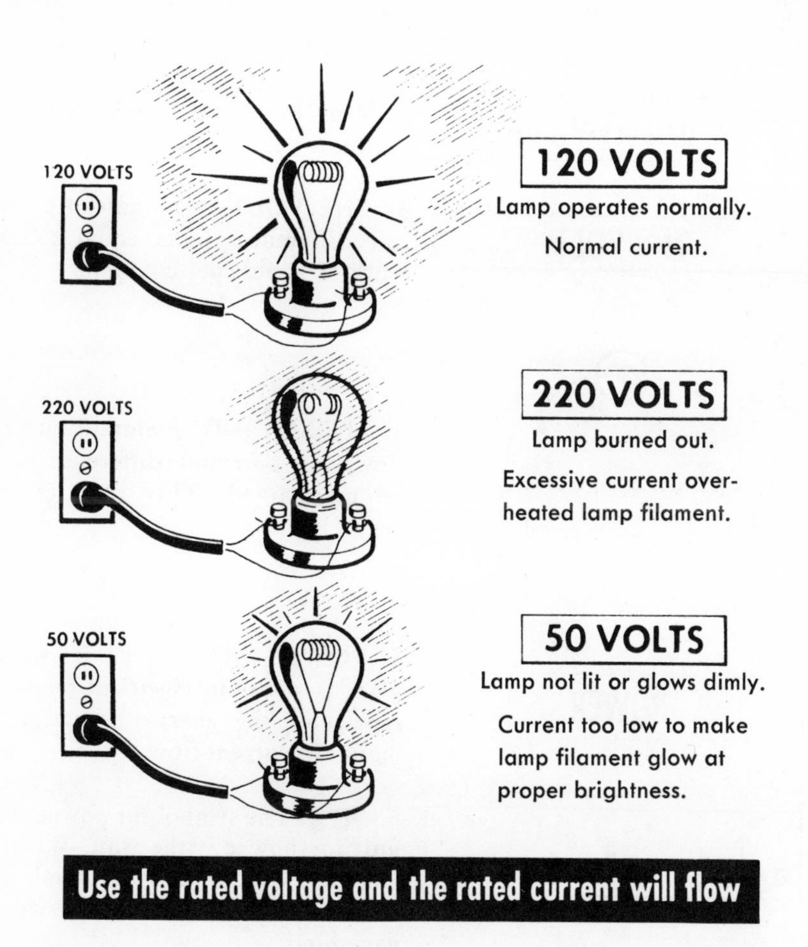

Use the rated voltage and the rated current will flow

Review of EMF or Voltage

To make current flow, a potential difference must be maintained between the terminals. When current flows, energy is required to maintain this difference and work must be done. Thus, the generation of electricity is the conversion of other forms of energy into this potential difference. At the device where the electricity is to be used, the potential difference makes the current flow and this is used to convert the electrical energy to some other form of energy or work. The rate at which this work is done is called *power*. We will say more about this in the next section.

EMF=Electromotive Force

1. EMF = ELECTROMOTIVE FORCE—The force that makes the current flow. Potential difference between terminals.

V or E=Voltage

2. V or E = VOLTAGE—Symbols used to designate the emf.

The VOLT

3. THE VOLT—The unit of potential difference. It is equal to the work of 1 joule per coulomb.

4. ENERGY—To maintain current flow, the potential difference must be maintained. This requires energy.

POWER

5. POWER—The rate at which work is done. In electrical terms, it represents the energy necessary to maintain current flow.

P

6. "P"—The symbol for power. The unit for power is the watt. When 1 ampere flows with a potential difference of 1 volt, 1 watt of power is generated.

Self-Test—Review Questions

. What is emf?
. What quantity is used to designate the magnitude of emf?
. What are the symbols used to designate emf?
. Define the unit of potential difference.
. What is the essential difference between current flow between two charged bars and that across the terminals of a battery?
. What is needed to maintain a potential difference between terminals where current is flowing?
. What happens if the potential difference is not maintained?
. What happens in a circuit where the voltage is too low? Why?
. What happens in a circuit where the voltage is too high? Why?
. What happens when the rated voltage is applied? Explain.
. What is electrical power?
. What is the symbol for power?
. Define the watt as the unit of power.
. Name some sources for energy that can be used to generate power.

Learning Objectives—Next Section

Overview—Now that you know something about what electricity is, you will study how electricity is produced and the many uses we have for electricity.

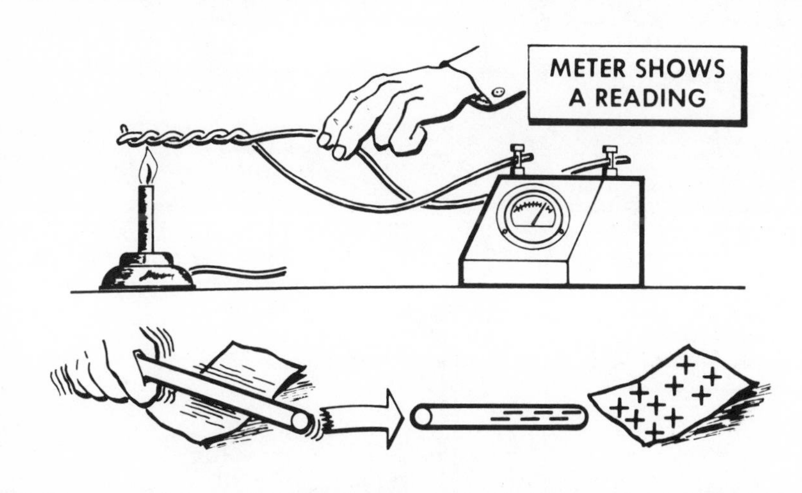

Electricity Is the Means for Transporting Power

You have learned about the electronic nature of matter and how electricity is the flow of electrons from place to place or the accumulation of electrons on a charged body. It is apparent that most materials are electrically balanced. What is needed is a source of *external* energy so that excesses and/or deficiencies of electrons can be maintained when current flows. By supplying external energy in the right way to the right device, we can generate electricity. On the other hand, by supplying electricity to the right devices, we can convert the energy of electricity to other useful forms such as mechanical power from motors, heat from appliances, light from lamps, etc. Thus, *electricity can be considered as the means for the convenient transportation and distribution of power*. For example, the energy from a waterfall can be harnessed to a generator to make electric power that is transported by transmission lines for hundreds of miles to a city where it is used to provide mechanical power, light, heat, cooling, and other necessities. In essence, then, we are using the power of the falling water when we use this electricity.

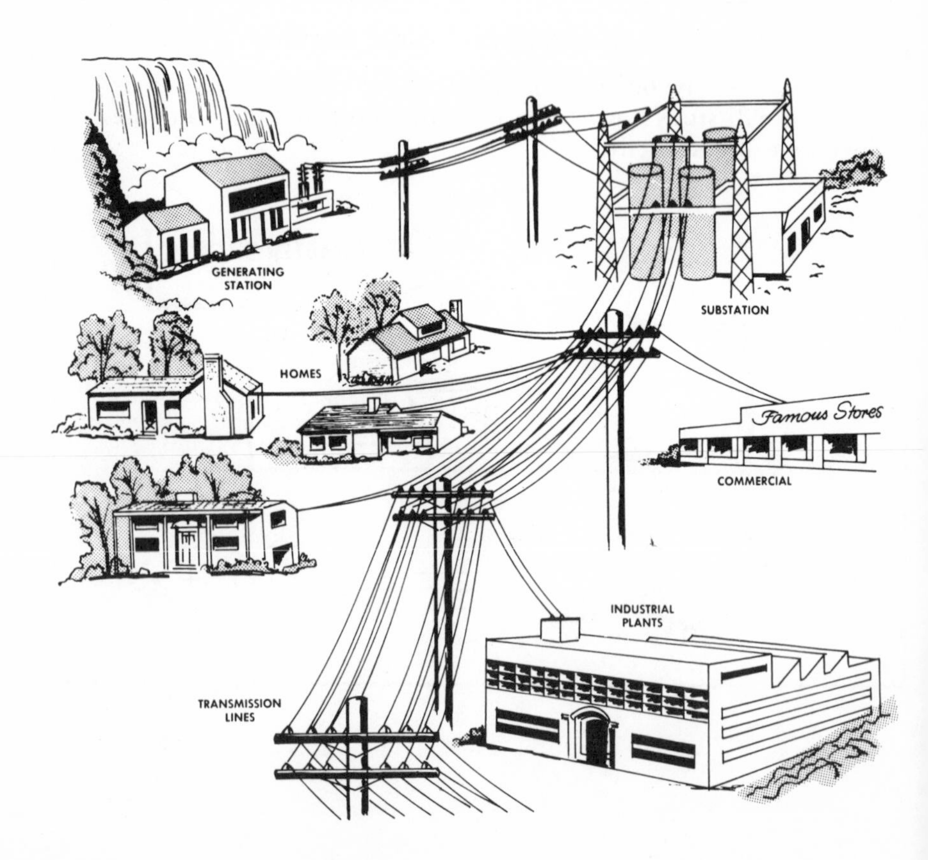

How Electricity Is Produced

The most common source of electricity and, in fact, the source for almost all the electrical energy that we use is electricity obtained from the interaction of conductors with magnetic fields. The second most common source involves chemical action and the example that you are most likely to see is the battery. Other energy sources for generating electricity in decreasing order of importance are light, heat, pressure, and friction. Remember, the generation of electricity is the conversion of other forms of energy into the potential difference required to make current flow.

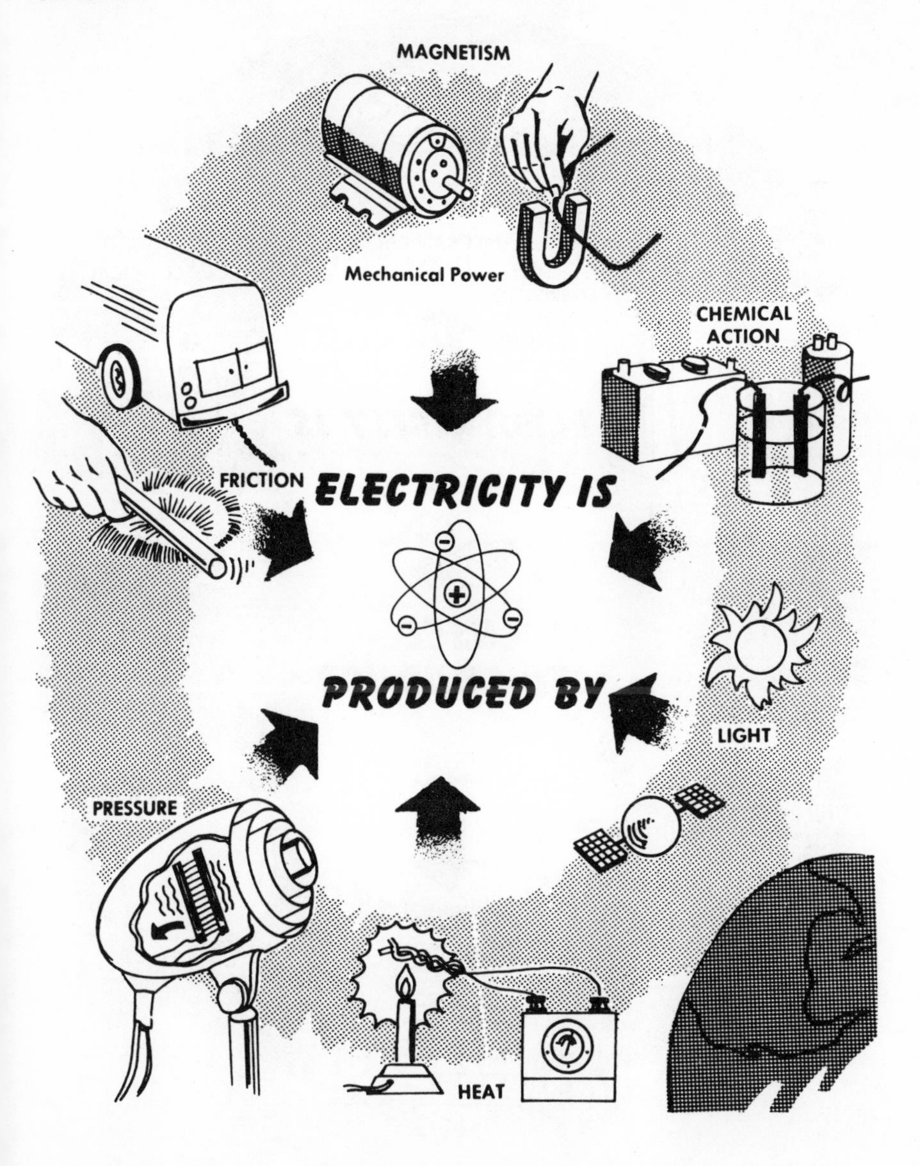

Uses of Electricity

Electricity, in turn, can be used to produce the very same effects that originally were utilized to produce that electricity with the exception of friction. These common uses of electricity (in reverse order) are mechanical power from motors, chemical action, light, heat, and pressure; and to operate electronic devices which will not be considered here.

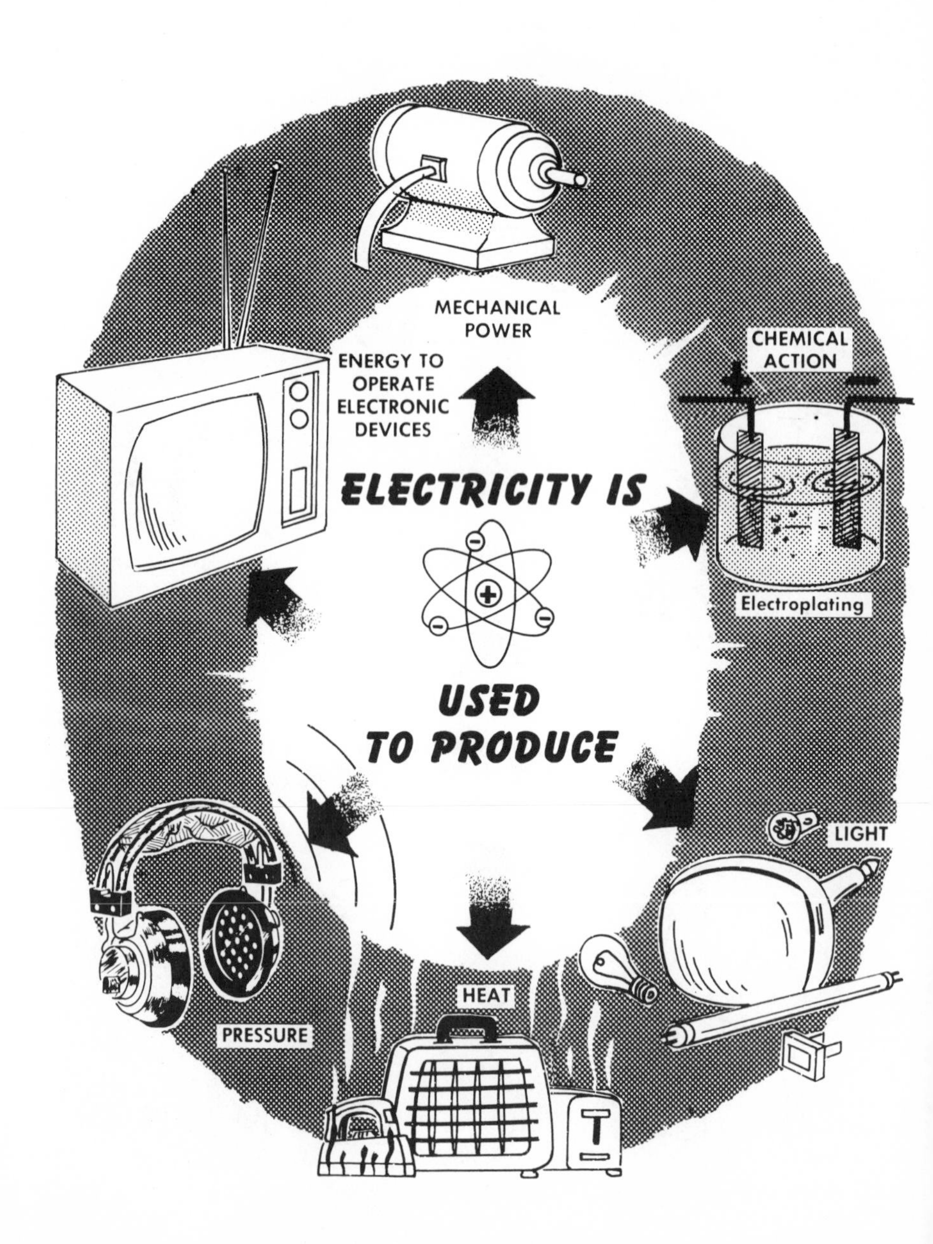

Electricity Produced from Frictional Energy (Static Electricity)

Although frictional electricity is the least important of all methods for the production of electricity, it is of value to study since it does have some useful applications and is important in understanding electric charges. You learned earlier that static (or frictional) electricity can be produced by rubbing certain dissimilar materials together. The source of energy in this case is from the muscles in your arm that, in turn, causes the separation of the charges. One of the applications of frictional or static electricity is in a device used in atomic research called the Van de Graf generator that will generate miniature bolts of lightning. Frictional electricity is usually a nuisance as mentioned earlier.

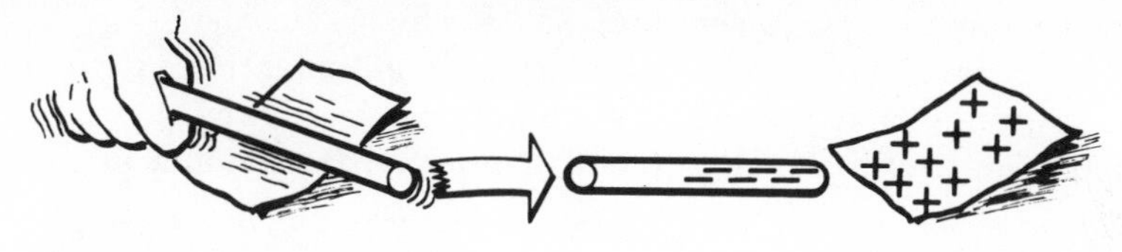

Although now we do not usually generate static electricity for use by friction, static electricity has some important applications. A most important application is the use of static electricity in electrostatic precipitators to remove carbon, fly ash, and other particles from the gases leaving a smokestack. This is done by giving the particles a charge of one polarity as they start to pass up the stack and then attracting them to collecting surfaces with a charge of opposite polarity further up the stack. In this way, most of the solid material present in the smoke can be removed. Techniques like this are most important for control of air pollution from industrial plants.

Another important application of static electricity is its use in the development of the xerographic copier—Xerox, IBM, etc. Electrostatic effects are utilized in the techniques of printing copies of the originals being duplicated.

Another useful application of static electricity is in electrostatic painting procedures (developed by the Ransburg Corporation) to paint on a high-volume, assembly-line basis even such irregularly shaped objects as automobile bodies, refrigerator cabinets, etc.

Electricity Produced from Pressure/Pressure Produced from Electricity

Electricity produced from pressure is called *piezoelectricity*, which is produced by certain crystalline materials. Crystals are orderly arrays of atoms in contrast to noncrystalline materials that have their atoms in a random pattern. Many crystalline materials exist in nature and many more can be made in the laboratory. Crystalline materials can be either pure elements or can be a compound. Two examples of naturally occurring crystals are quartz, which is the major component of common beach sand, and the diamond, which is the crystalline form of carbon. The most common materials used for the production of piezoelectricity are quartz, barium titanate (a ceramic), and Rochelle salts.

If a crystal made of these materials is placed between two metal plates and pressure is exerted on the plates, an electric charge will be developed. The size of the charge will depend on the amount of pressure exerted.

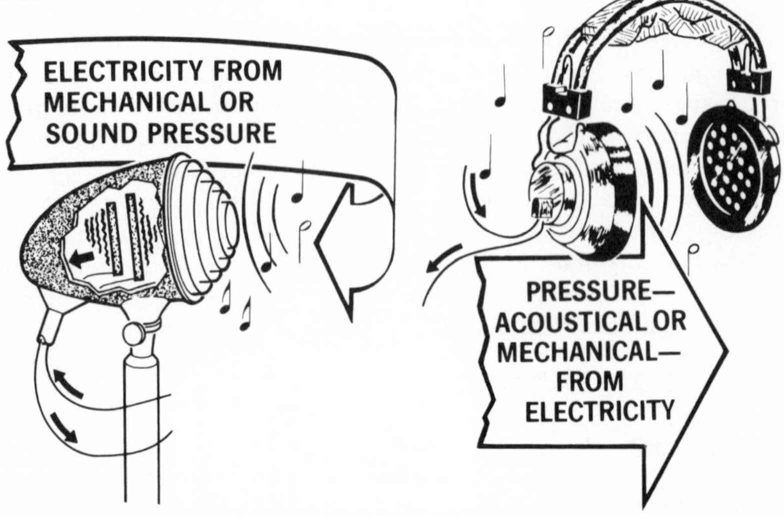

While the actual use of pressure as a source of electricity (piezoelectricity) is limited to very low power applications, you will find it in many different kinds of equipment. Crystal microphones, crystal phonograph pickups, and sonar equipment use crystals to generate electric charges from pressure. In these applications the mechanical energy comes from sound pressure or acoustical energy that moves a diaphragm that is mechanically coupled to the crystal; or, in the case of the phonograph pickup, the motion of the needle is coupled to the crystal. *Conversely,* if an electric charge is placed across the metal plates, the crystal will distort or physically change its shape, generating acoustical or mechanical energy. This is the principle used in crystal headphones. These sources and uses are entirely involved with electronic equipment and will not be further discussed in our study of electricity.

Electricity Produced from Heat (and Cold)

If a length of metal, such as copper, is heated at one end, electrons tend to move away from the hot end toward the cooler end. While this is true of most materials, some, such as iron, work the other way; that is, the electrons tend to flow toward the hot end. Thus, if an iron wire and a copper wire are twisted together to form a junction and the junction is heated, the flow of electrons will result in a charge difference between the free ends of the wires. (It should also be mentioned that if *cold* is applied, electrons will flow, but in the *opposite* direction.) In the illustration, the energy is supplied as heat from the burner.

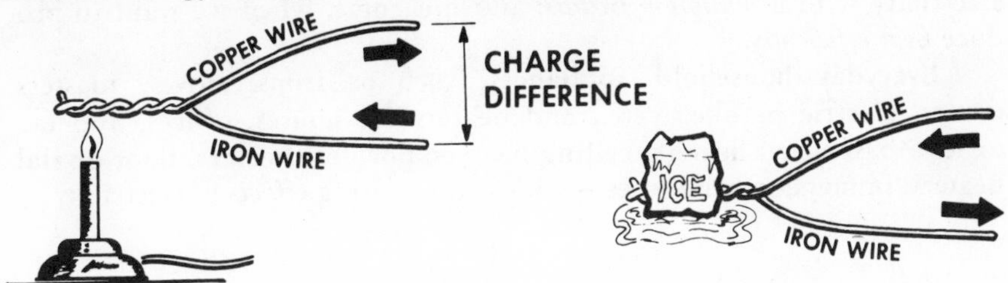

The amount of charge produced depends on the difference in temperature between the junction and the opposite ends of the two wires. A greater temperature difference results in a greater charge.

A junction of this type is called a *thermocouple* and will produce electricity as long as heat is applied. While twisted wires may form a thermocouple, more efficient thermocouples are constructed of two pieces of dissimilar metal riveted or welded together.

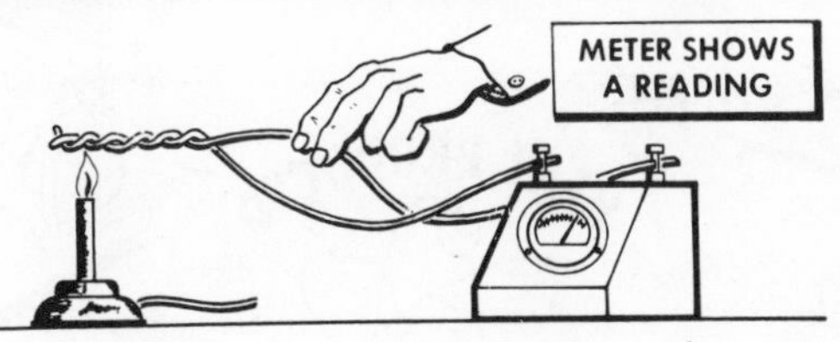

Since the current flow is proportional to the temperature of the junction, a thermocouple can be used to measure temperature when connected to a suitable indicating device. It is often used for this purpose, for example in an automobile engine temperature indicator. Many thermocouples can be connected together to form a *thermopile* or *thermoelectric generator*. Thermopiles are used in measuring heat inside high-temperature furnaces; also, in the fail-safe pilot flame thermal device in home gas furnaces to shut off the gas valve in case the pilot goes out. These generators using semiconductor materials are becoming more common as a replacement for batteries, particularly for military application. Their major advantage is that they will deliver power as long as the heat source is turned on. Their major disadvantages are that they are not capable of delivering very much power even in the larger sizes and are not very efficient.

Heat Produced from Electricity

Some heat is always produced when an electric current flows through a wire that is not a perfect conductor. This is because some energy is lost or used up—in the form of heat—in causing the electric current to flow. Good conductors produce less heat—although always some—because it is easy to cause current to flow through them. Poor conductors—as, for example, nichrome, an alloy of nickel and chromium used to make heating elements—produce a great deal of heat when current flows through them. Copper, for example, is about 60 times as good a conductor as nichrome. Thus, we use copper when we want to deliver electricity with a *minimum* of *loss*; and nichrome, when we want to produce *heat efficiently*.

Everyday household appliances, such as irons, stoves, toasters, dryers, electric blankets, etc., and heaters for houses, offices, and factories—baseboard heaters, ceiling heaters, portable heaters, floor or slab heaters, immersion heaters, etc—all use the heating effects of electricity.

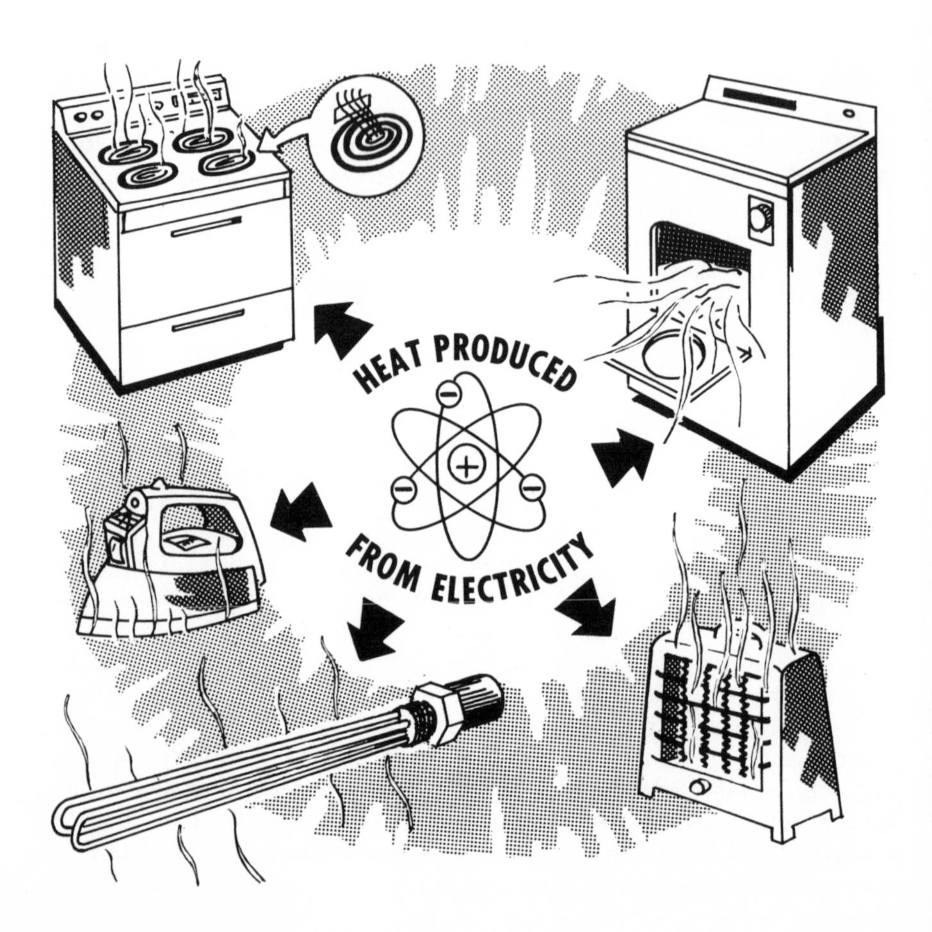

Electricity Produced from Light

Electricity may be produced (or controlled) by using light as a source of energy. This occurs because materials like potassium, sodium, germanium, cadmium, cesium, selenium, and silicon release electrons when excited by light under the right conditions. This release of electrons is called the *photoelectric effect.*

The photoelectric effect is used in three ways. Light causes *photoemission*; the incident beam of light causes a surface to emit electrons that are collected to form an electric current. Light causes a change in how well a material conducts electricity. This is called the *photoconductive effect.* The third effect is the *photovoltaic effect*; the energy from the incident beam of light is converted directly into a flow of electrons. Although the photoconductive and photoemissive effects are very useful, particularly in electronic systems, the only source for significant amounts of electricity from light involves the principle of the photovoltaic effect. The solar cells that are used to power space vehicles and satellites are of this type. Let's take a look at how the photovoltaic cell is made and how it works.

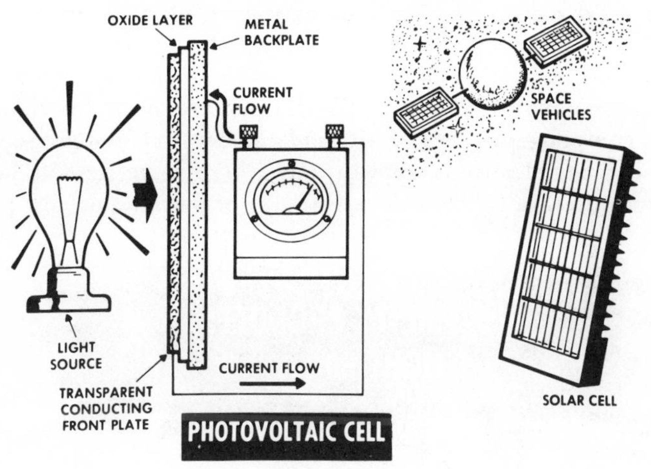

Most photovoltaic cells are made with selenium or silicon as the basic material. The cell is made up as a three-layer sandwich consisting of a pure material backing plate (copper) covered with a layer of oxide (selenium or silicon dioxide) that forms the center of the sandwich and a very thin transparent or translucent conducting layer as the other side of the sandwich. Light releases electrons from the junction of the oxide and the front plate. These electrons flow through the external circuit and back to the back plate. The electrical energy from each cell is very small, but if the cells are made larger to intercept more light, or many cells are connected together, then a significant amount of power can be developed. At least, there can be enough generated to power electronic equipment in satellites and similar space vehicles.

Light Produced from Electricity

Many of the poorer conductor materials, such as tungsten, glow red and even white hot when they become heated from conducting electric current. This radiant glow—incandescence—gives off *light* as well as heat. The incandescent lamp works this way to produce much of the light that we use.

Light is also produced with less electricity and without much heat by *fluorescence, phosphorescence* and *electroluminescence.* The fluorescent lights used in houses, offices, factories, etc. contain a gas (argon, mercury vapor, etc.) which, when forced to conduct an electric current, becomes ionized and produces ultraviolet and some visible radiation. When this radiation strikes a fluorescent coating inside the fluorescent light tube, it gives off a colored light. By using a mixture of fluorescent materials of different colors, white light can be produced.

Television picture tubes work on the principle of *phosphorescence* in which an electron beam strikes a surface coated with phosphorescent material which, in turn, gives off light.

Some light sources work on the principle of *electroluminescence* in which a solid material, when conducting an electric current, gives off light. This is the general principle by which the electroluminescent or light-emitting diode, LED (typically made of gallium arsenide phosphide wafers) works. These LED devices are used in digital wristwatches, computer displays of all kinds, pocket calculators, security systems, and other electronic applications where some light is needed for special purposes. LEDs have the advantage of working on low power.

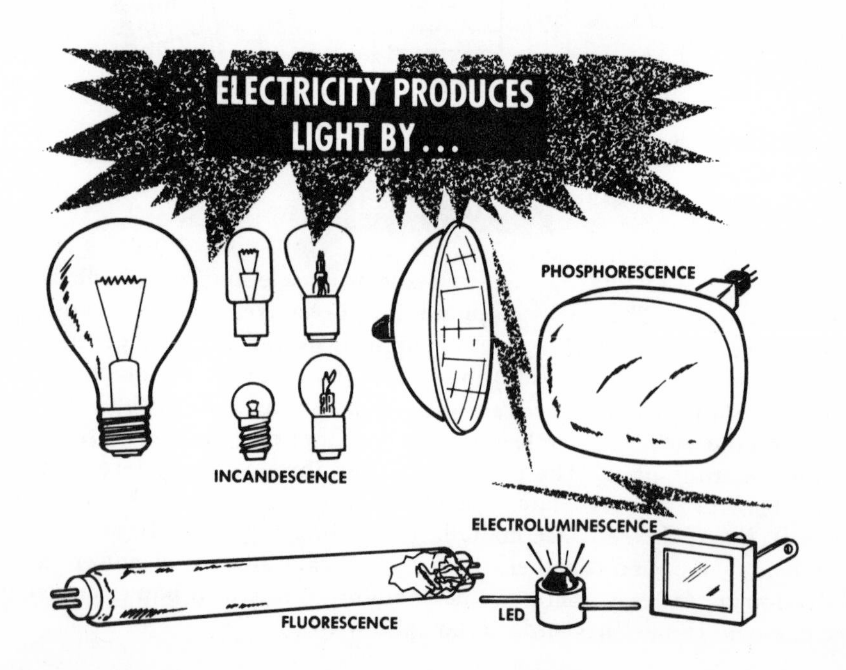

Electricity Produced from Chemical Action

Batteries are the most common source of electricity from chemical action. A *fuel cell* is another device used to generate electricity by chemical action. Fuel cells have the advantage of being light and capable of long life but are extremely expensive and, therefore, are only used for military and spacecraft applications at the present time. In the fuel cell, gases like hydrogen and oxygen are made to combine directly to form water and the energy released by this reaction is used directly to generate electricity. For most applications requiring a portable or emergency power source, batteries using chemical energy to generate electricity are used. We will discuss batteries only briefly here, but you will study them in much more detail later when you have learned more about electricity.

A battery is often made up of a number of identical cells connected together in a common container. We hook up cells this way so we can get more energy than we could from a single cell. Thus, the cell is the fundamental unit of the battery. When you learn about cells, you will know about batteries. Although materials may differ for different batteries, all batteries consist of two dissimilar metal plates called *electrodes* immersed in an *electrolyte* that may be either a paste or a liquid.

You will encounter batteries in many electric and electronic circuits. When reading diagrams showing how these circuits are hooked up, you will find that the various circuit elements are represented by a special representation called the *schematic symbol*. The schematic symbols shown above are used for all batteries. It is not necessary to show all of the cells in the schematic representation if we write adjacent to the battery the battery output potential, such as 6 volts.

Electricity Produced from Chemical Action (continued)

All cells consist of two electrodes and an electrolyte. Electricity is produced by the chemical reactions that take place between the electrodes and the electrolyte. The simplest cell—a *wet cell*—consists of strips of zinc metal and copper metal for the electrodes and an acid solution, for example sulfuric acid and water, as the electrolyte. In a *dry cell*, such as a flashlight battery, the electrolyte is in the form of a paste, rather than a fluid.

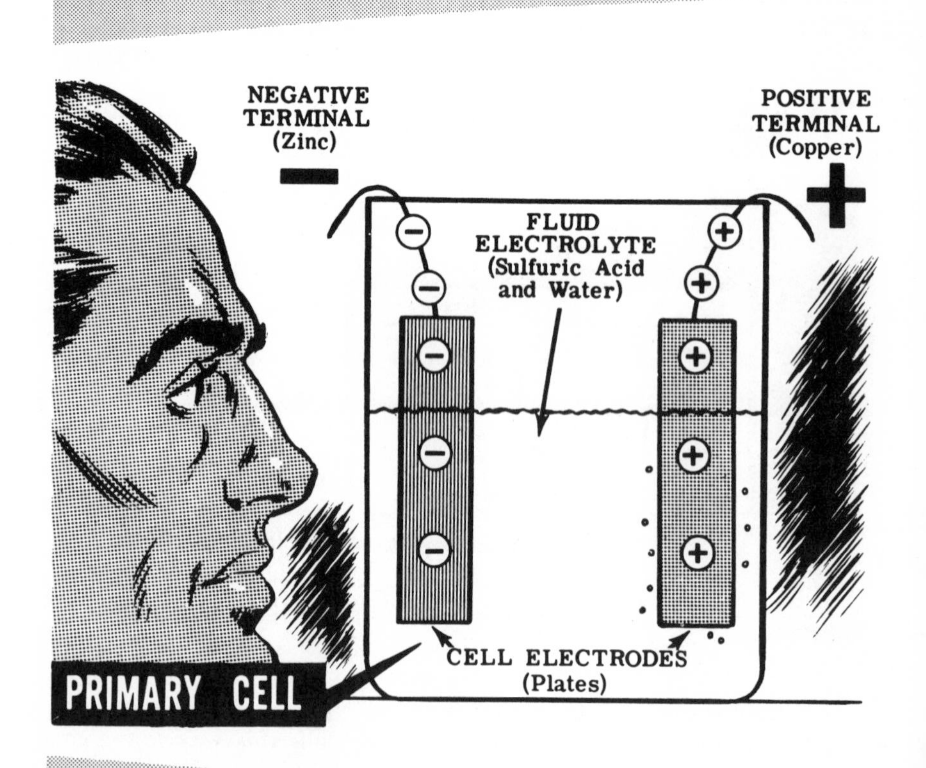

We will discuss the details of how electricity is obtained by chemical action later. In the cell above, the zinc is very slowly dissolved in the electrolyte. As the zinc atoms go into the solution they leave electrons on the undissolved zinc electrode. Thus, the zinc electrode develops a negative charge. By a similar process, electrons leave the copper electrode to unite with hydrogen atoms from the sulfuric acid electrolyte to form neutral hydrogen atoms and a positive charge develops on the electrode. The copper electrode is not dissolved since it supplies only electrons.

Electricity Produced from Chemical Action (continued)

With nothing connected to the cell terminals, you would see that electrons are pushed onto the negative plate until there is room for no more. The electrolyte would take enough electrons from the positive plate to make up for those it pushed onto the negative plate. Both plates would then be fully charged and no electrons would be moving between the plates.

Now suppose you connected a wire between the negative and positive terminals of the cell. You would see the electrons on the negative terminal leave the terminal and travel through the wire to the positive terminal. Since there would now be more room on the negative terminal, the electrolyte would carry more electrons across from the positive plate to the negative plate. As long as electrons leave the negative terminal and travel to the positive terminal outside the cell, the electrolyte will carry electrons from the positive plate to the negative plate inside the cell.

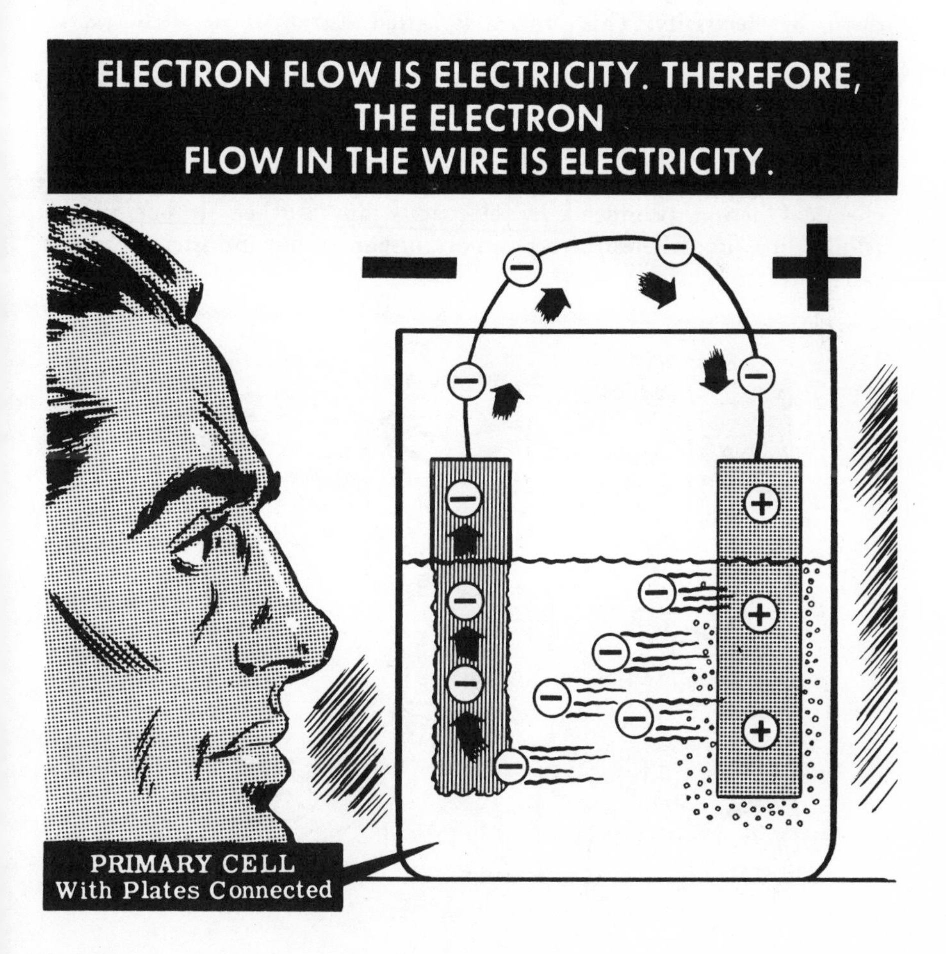

Chemical Action Produced from Electricity

Probably the most common example of chemical action produced from electricity is the recharging of the ordinary automobile storage battery. When the cells of the storage battery are being used to generate electricity, a chemical reaction takes place. If a current is sent through the cells in the opposite direction, the reaction runs in the other direction and the battery is recharged. Cells that do this are called *secondary cells*. Most secondary cells used in storage batteries are of the lead-acid type. In this cell, the electrolyte is sulfuric acid, the positive plate is lead peroxide, and the negative plate is lead. During discharge of the cell, the acid becomes weaker and both plates change chemically to lead sulfate. Recharging reconverts the lead sulfate to pure lead on one plate and lead peroxide on the other, and the strength of the sulfuric acid electrolyte increases. Other types of secondary cells use nickel and iron, nickel and cadmium, or silver and zinc in a potassium hydroxide electrolyte.

Since the basic force that holds compounds together is electrical in nature, it is not surprising that chemical compounds can be broken down by electricity. This process is called *electrolysis*, or *electrolytic action*, and is very important in the manufacture of many metals (aluminum, copper, etc.) and other substances. An additional important use of chemical action produced from electricity is in *electroplating*. Here, metal ions are made to migrate to an electrode and adhere to it when they are changed from ions to the metal. Although we will not look into the chemical action produced by electricity any further in our present studies, it is nonetheless a very important part of our industrial and personal life.

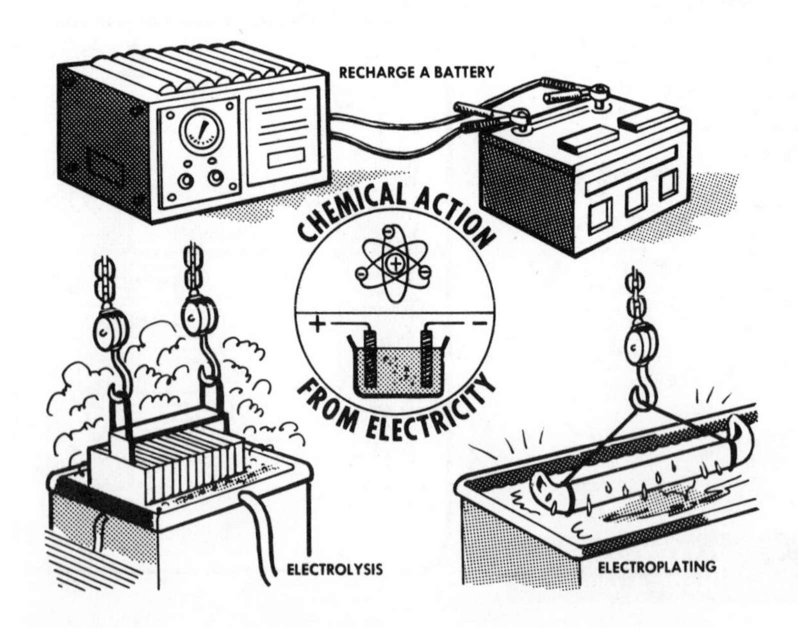

Electricity Produced from Magnetism

The most common method of producing electricity for electric power is by the use of magnetism. The source of electricity must be able to maintain a large potential difference because the charge is being used to furnish electric power.

Almost all of the electric power used, except for emergency and portable equipment operated from batteries, originally comes from a generator in a power plant. The generator may be driven by water power, a steam turbine with its steam heated by coal, oil, gas, or atomic power, or an internal combustion engine. No matter how the generator is driven, the electric power it produces is the result of the action between the wires and the magnets inside the generator. Remember, electricity does not by itself produce power, it transports it.

When wires move past a magnet or a magnet moves past wires, electricity is produced in the wires. Now you will find out how magnetism is used to produce electricity.

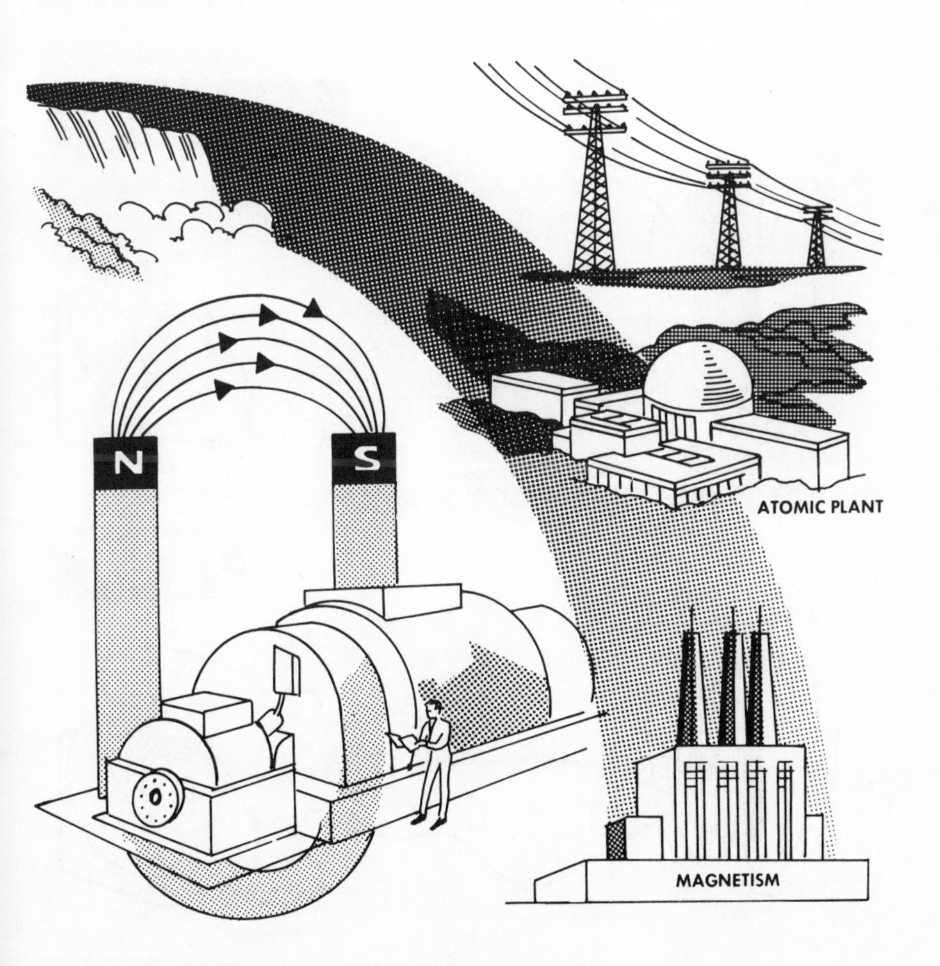

Electricity Produced from Magnetism (continued)

One method by which magnetism produces electricity is through the movement of a magnet past a stationary wire. If you connect a very sensitive meter (a device for indicating current flow) across the ends of a stationary wire and then move a magnet past the wire, the meter needle will deflect. This deflection indicates that electricity is produced in the wire. Repeating the movement and observing the meter closely, you will see that the meter moves only while the magnet is passing *near* the wire.

Placing the magnet near the wire and holding it at rest, you will observe no deflection of the meter. Moving the magnet from this position, however, does cause the meter to deflect and shows that, alone, the magnet and wire are not able to produce electricity. In order to deflect the needle, *movement* of the magnet past the wire is necessary.

Movement is necessary because the magnetic field, or flux lines, around a magnet produces an electric current in a wire only when the magnetic field is moved across the wire. When the magnet and its field are stationary, the field is not moving across the wire and will not produce a movement of electrons in it.

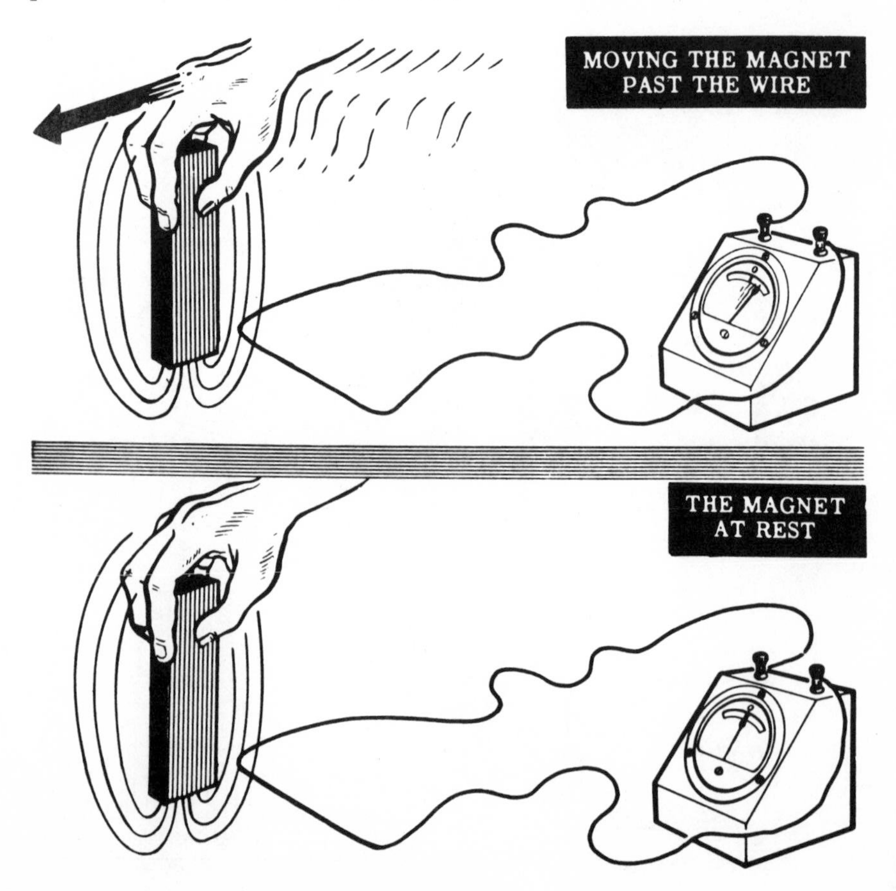

Electricity Produced from Magnetism (continued)

In studying the effects of moving a magnet past a wire, you discovered that electricity was produced only while the magnet and its field were actually moving. If you move the wire past a stationary magnet, you again will notice a deflection of the meter. This deflection will occur only while the wire is moving across the magnetic field.

To use magnetism to produce electricity, you may either move a magnetic field across a wire, or move a wire across a magnetic field. In either case, it is the wire cutting across the lines of force or flux lines that produces electricity. For a continuous source of electricity, however, you need to maintain a *continuous* motion of either the wire or the magnetic field.

To provide a continuous motion, the wire or the magnet would need to move back and forth constantly. A more practical way is to cause the wire to travel in a circle through the magnetic field.

This method of producing electricity—that of the wire traveling in a circle past the magnets—is the principle of the *electric generator* (dynamo) and is the source of most electricity used for electric power.

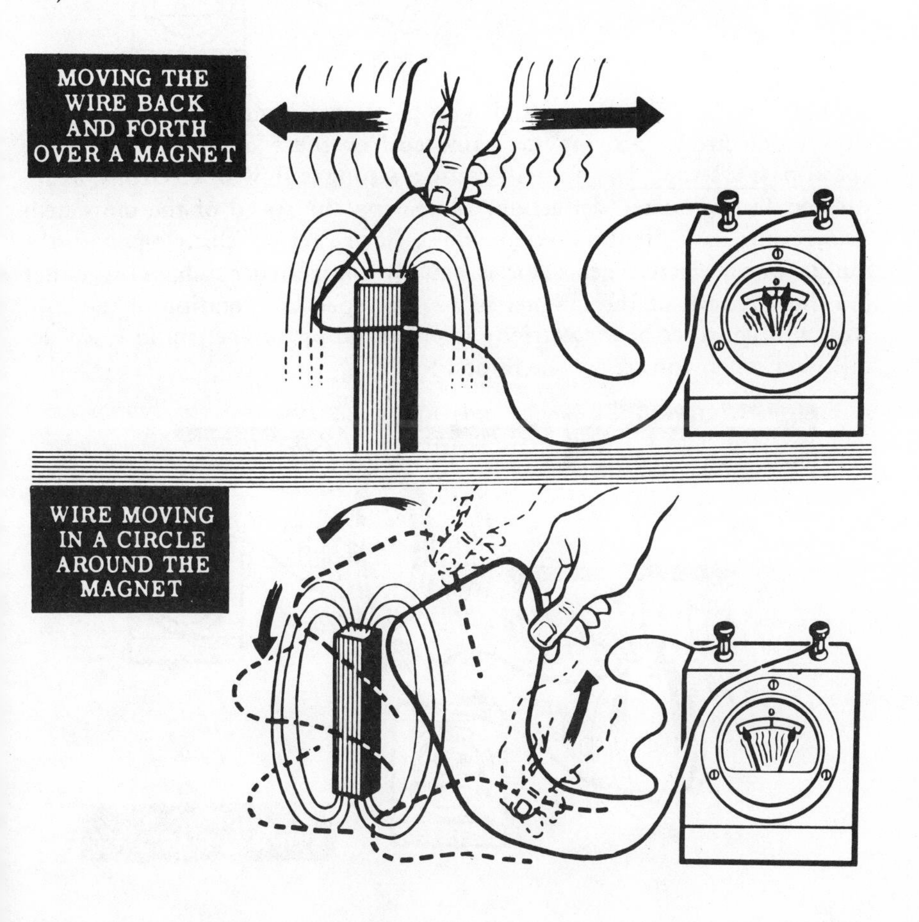

Electricity Produced from Magnetism (continued)

Since the electricity is produced by the wire cutting past the flux lines, you can change the *amount* of electricity produced by changing the strength of the magnetic field or by cutting more flux lines with the wire in a shorter length of time.

To increase the amount of electricity produced by moving a wire past a magnet, you might increase the length of the wire that passes through the magnetic field, use a stronger magnet, or move the wire faster. The length of the wire can be increased by winding it in several turns to form a *coil*. Moving the coil past the magnet will result in a much greater deflection of the meter than resulted with a single wire. Each additional coil turn will add an amount equal to that of one wire.

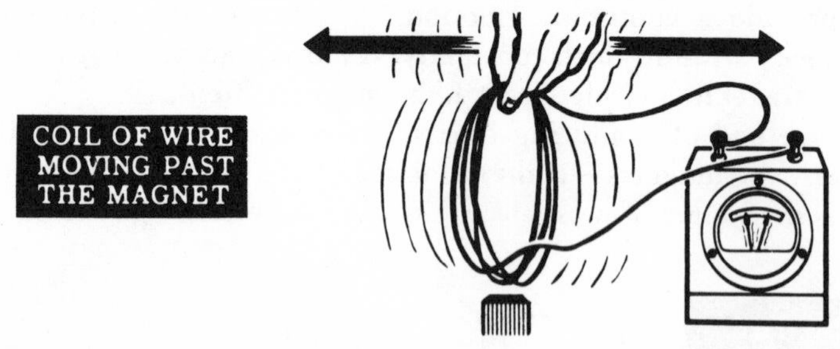

Moving a coil or a piece of wire past a weak magnet causes a weak flow of electrons. Moving the same coil or piece of wire at the same speed past a strong magnet will cause a stronger flow of electrons, as indicated by the meter deflection. Increasing the speed of the movement also results in a greater electron flow. In producing electric power, the output of an electric generator is usually controlled by changing either (1) the strength of the magnet or (2) the speed of rotation of the coil. You will consider how electricity is produced by magnetism in great detail later when you study generators.

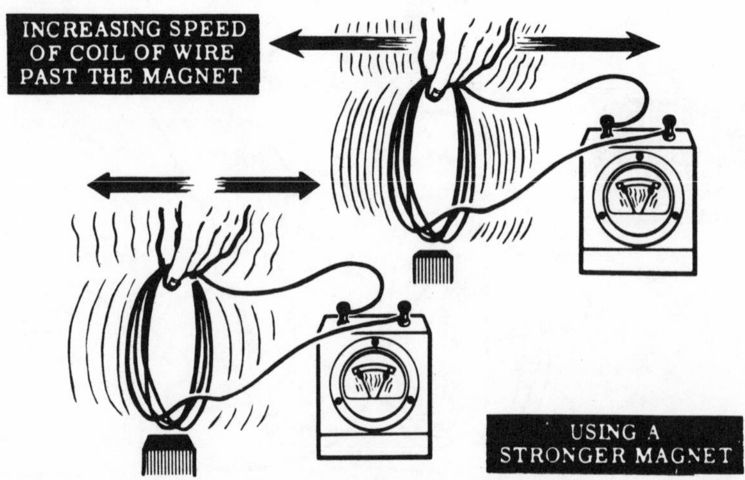

Magnetism Produced from Electricity

Magnetic fields can be created conversely, by electricity, just as you might suspect, because you know that you can produce electricity from magnetism. Any conductor that carries a current will act like and is, in fact, a magnet. If the wire is wound into a coil, the magnet will be stronger. The magnet will be stronger also if the current is increased. Since the magnetic field depends on the flow of current, there is no magnetic field if the current is removed. Temporary magnets of this type are called *electromagnets* and the effect is called *electromagnetism*. Electromagnetism is so important in your study of electricity that the entire next section is devoted to that subject.

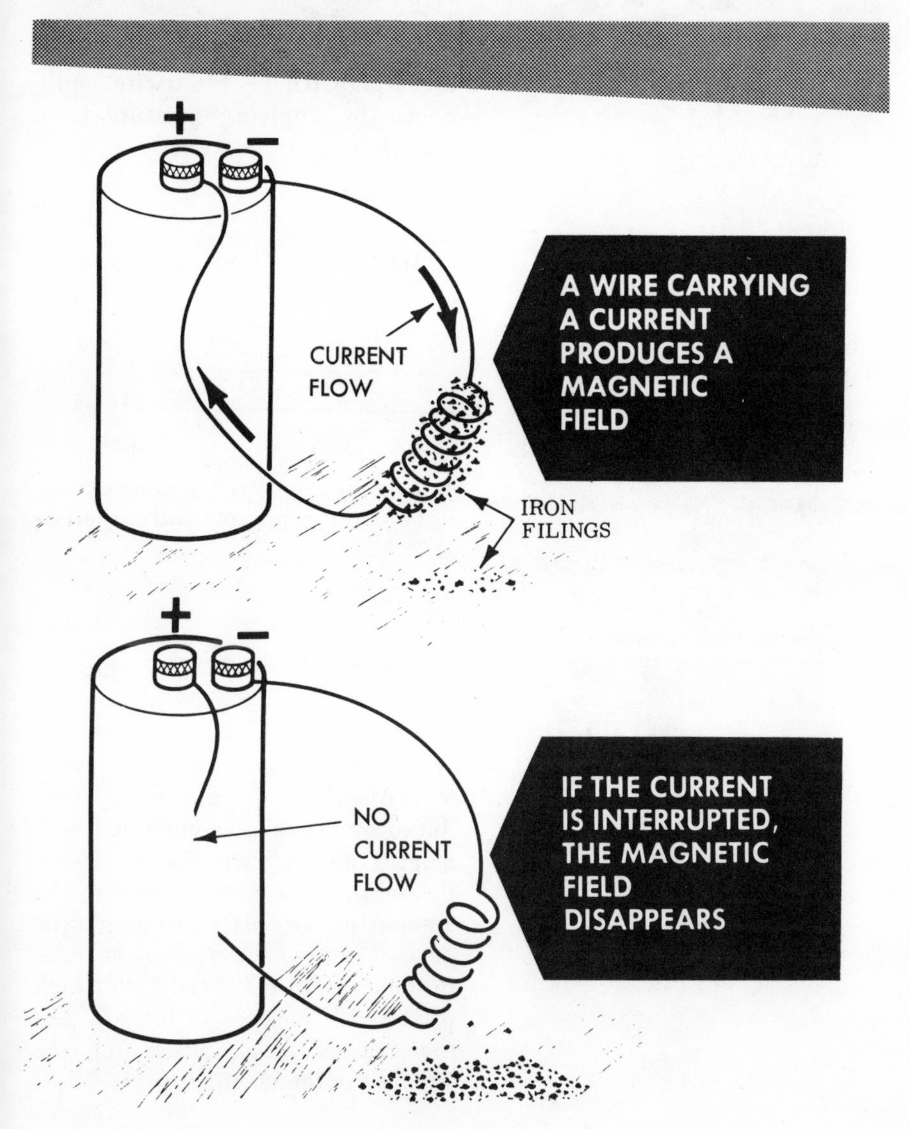

Review of How Electricity Is Produced

Electricity is the action of electrons which have been forced from their normal orbits around the nucleus of an atom. To force electrons out of their orbits so they can become a source of electricity, some kind of energy is required.

Six kinds of energy can be used:

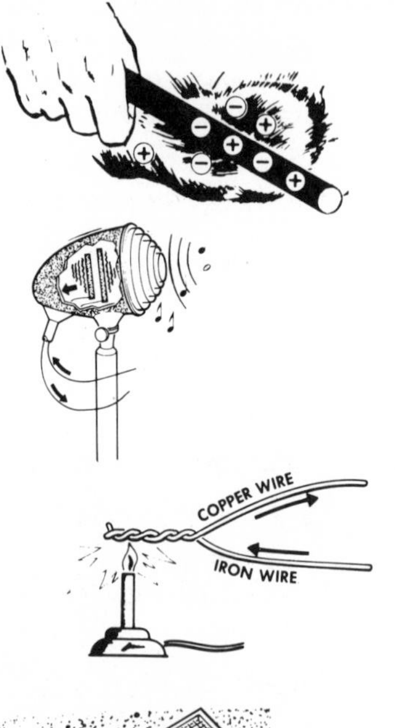

1. FRICTION — Electricity produced by rubbing two materials together.

2. PRESSURE — Electricity produced by applying pressure to a crystal of certain materials.

3. HEAT—Electricity produced by heating the junction of a thermocouple.

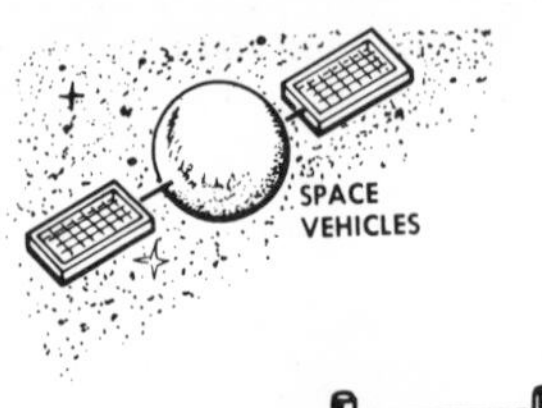

4. LIGHT—Electricity produced by light striking photosensitive materials.

5. CHEMICAL ACTION—Electricity produced by chemical reaction in an electric cell.

6. MAGNETISM—Electricity produced by the relative movement of a magnet and a wire that results in the cutting of lines of force. The amount of electricity produced will depend on: (a) the number of turns in the coil; (b) the speed with which the relative motions of the coil and the magnet take place; (c) the strength of the magnet.

Review of How Electricity Is Used

Electricity, or the flow of current, is used by all of us to do a number of tasks in all aspects of our everyday life, as well as in some ways that are not so obvious. Usually when electricity is used, it is converted into some useful form of energy.

1. PRESSURE—If a potential difference is applied across the faces of certain kinds of crystals, such as Rochelle salt, the crystal will distort and produce pressure or mechanical movement. This is the principle used in crystal headphones.

2. HEAT—When current flows through an imperfect conductor, some of the energy is used up in getting the electrons through. This energy appears as heat. Poor conductors that do not melt easily, such as nichrome wire, are used as heating elements.

3. LIGHT—When enough current is passed through a wire, it can become white hot or incandescent. This is the way our ordinary light bulb works, and it produces light as well as heat. To keep the filament (heated wire) from burning up, the filament is enclosed in a bulb with an inert gas. Electricity can also produce light by *electroluminescence*, *phosphorescence*, and *fluorescence*.

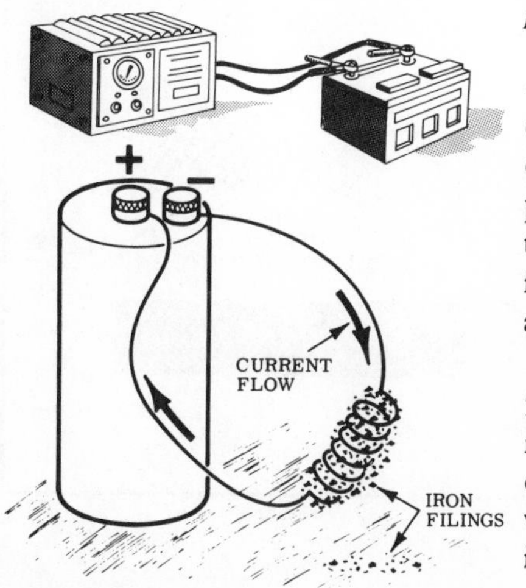

4. CHEMICAL ACTION—Electriccity can cause the decomposition of chemical compounds. This is the principle behind the secondary cells that are used in storage batteries. It is also the basis for electroplating and electrolytic action.

5. ELECTROMAGNETISM—Current passing through a wire produces a magnetic field around the wire as long as the current is flowing. This effect is *electromagnetism.*

Self-Test—Review Questions

1. What are the six common sources of electricity?
2. Which of these sources is most important? Least important? Why?
3. Describe how a thermocouple is made and how it operates.
4. What are the three necessary components for each cell of a battery?
5. How do primary and secondary cells differ? How are they alike?
6. What is the circuit schematic symbol for a cell? For a battery?
7. What is the difference between a cell and a battery?
8. Describe the basic principle involved with the generation of electricity from magnetism. What determines the amount of electricity produced?
9. How would you go about increasing or decreasing the amount of electricity produced by magnetism?
10. What is the common principle that is involved in the generation of electricity from any source?
11. What are the five major effects produced by electricity?
12. How does the incandescent lamp work?
13. Why does the flow of current in conductors produce heat?
14. What happens to iron filings placed near a wire or coil that is carrying current? Why?
15. What happens to the iron filings of Question 14 when the current is removed? Why?

Learning Objectives—Next Section

Overview—Earlier you learned about magnetism and some of the effects of magnetism. Now you will learn about a very important type of magnet called the *electromagnet*. Electromagnetism is one of the most important effects of electricity as you will see as you study further.

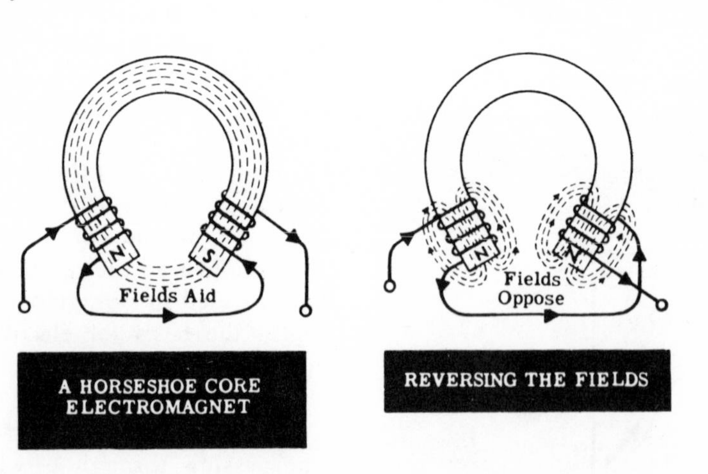

A HORSESHOE CORE ELECTROMAGNET

REVERSING THE FIELDS

Electromagnetism

Earlier you learned the very important fact that an electric current will flow when you move a coil of wire so that it cuts through a magnetic field. You also learned that this is the most widespread manner in which electricity is generated for the home, industry, aboard ship, etc. You learned too that electricity can generate magnetism. In this section you will see for yourself exactly how this is done.

Earlier you made use of permanent magnets to cause an electric current to flow. You saw that more current could be generated as you increased the number of turns of wire, the speed of motion of the coil, and the strength of the magnetic field. It is a simple matter to accomplish the first two of these in a practical electric generator, but it is very difficult to increase the strength of a permanent magnet beyond certain limits. In order to generate large amounts of electricity, a much stronger magnetic field must be used. This is accomplished by means of an *electromagnet*. Electromagnets work on the simple principle that a magnetic field is generated by passing an electric current through a coil of wire. As you learned, electromagnets differ from permanent magnets in that they are magnetic only when an electric current is supplied.

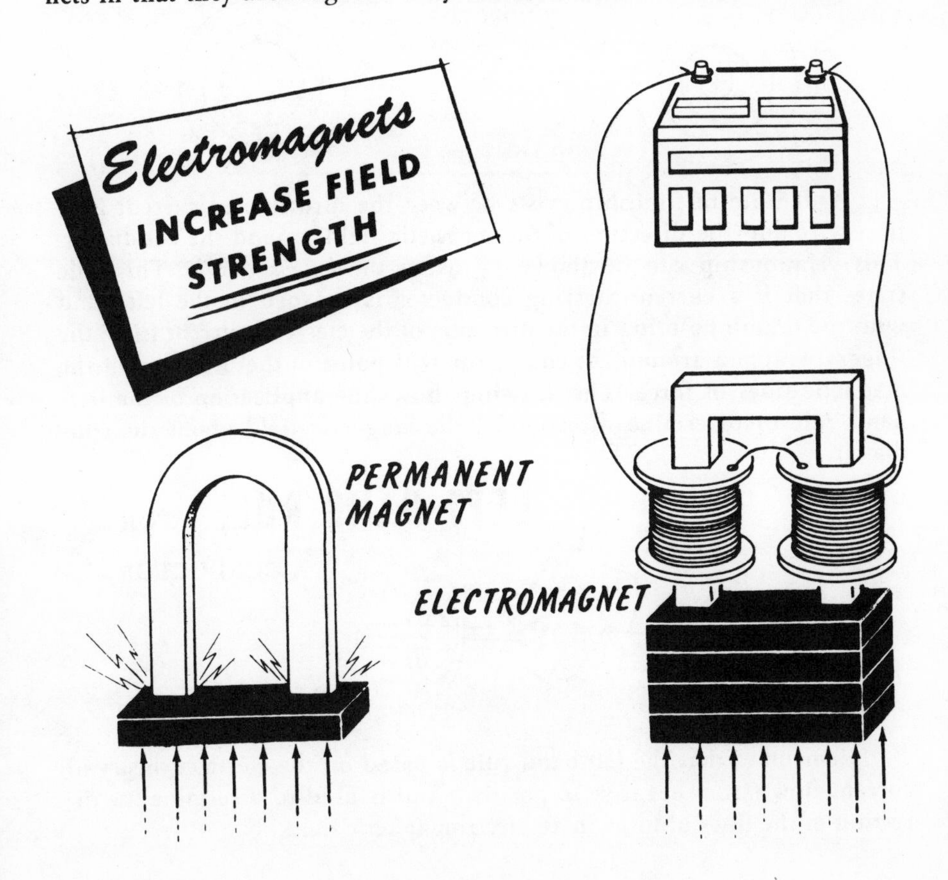

Magnetic Fields around a Conductor

An electromagnetic field is a magnetic field caused by the current flow in a wire. Whenever electric current flows, a magnetic field exists around the conductor, and the direction of this magnetic field depends upon the direction of current flow. The illustration shows conductors carrying current in different directions. The direction of the magnetic field is counterclockwise when current flows from left to right. If the direction of current flow *reverses*, the direction of the magnetic field also *reverses*, as shown. In the cross-sectional view of the magnetic field around the conductors, the dot in the center of the circle represents the head of the arrow indicating the current flowing out of the paper *toward* you; the cross represents the tail of the arrow indicating the current flowing into the paper *away* from you.

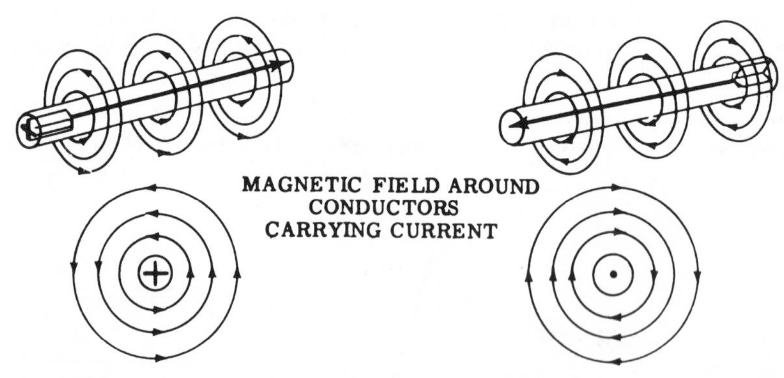

A definite relationship exists between the direction of current flow in a wire and the direction of the magnetic field around the conductor. This relationship can be shown by using the *left-hand rule*. This rule states that if a current-carrying conductor is grasped in the left hand with the thumb pointing in the direction of the electron current flow, the fingers wrapped around the conductor will point in the direction of the magnetic lines of force. The drawing shows the application of the left-hand rule to determine direction of the magnetic field about the conductor.

Remember that the left-hand rule is based on the electron theory of current flow (from negative to positive) and is used to determine the direction of the lines of force in an electromagnetic field.

Magnetic Fields around a Conductor (continued)

It is easy to demonstrate in an experiment that a magnetic field exists around a current-carrying conductor. Connect a heavy copper wire in series with a switch and a dry cell battery. The copper wire is bent to support itself vertically and then inserted through a hole in a lucite plastic sheet, which is held in a horizontal position. When the switch is closed, iron filings—which have the property of aligning themselves along the lines of force in a magnetic field—are sprinkled on the lucite. The lucite is tapped lightly to make it easier for the iron filings to fall into position.

If you did this experiment, you would see that the filings arrange themselves in concentric circles, showing that the magnetic lines of force form a circular pattern around the conductor. To show that the circular pattern is actually the result of the magnetic field, you could open the switch and spread the filings evenly over the lucite, then repeat the experiment. Each time the circuit current flows, the filings arrange themselves to show the magnetic field.

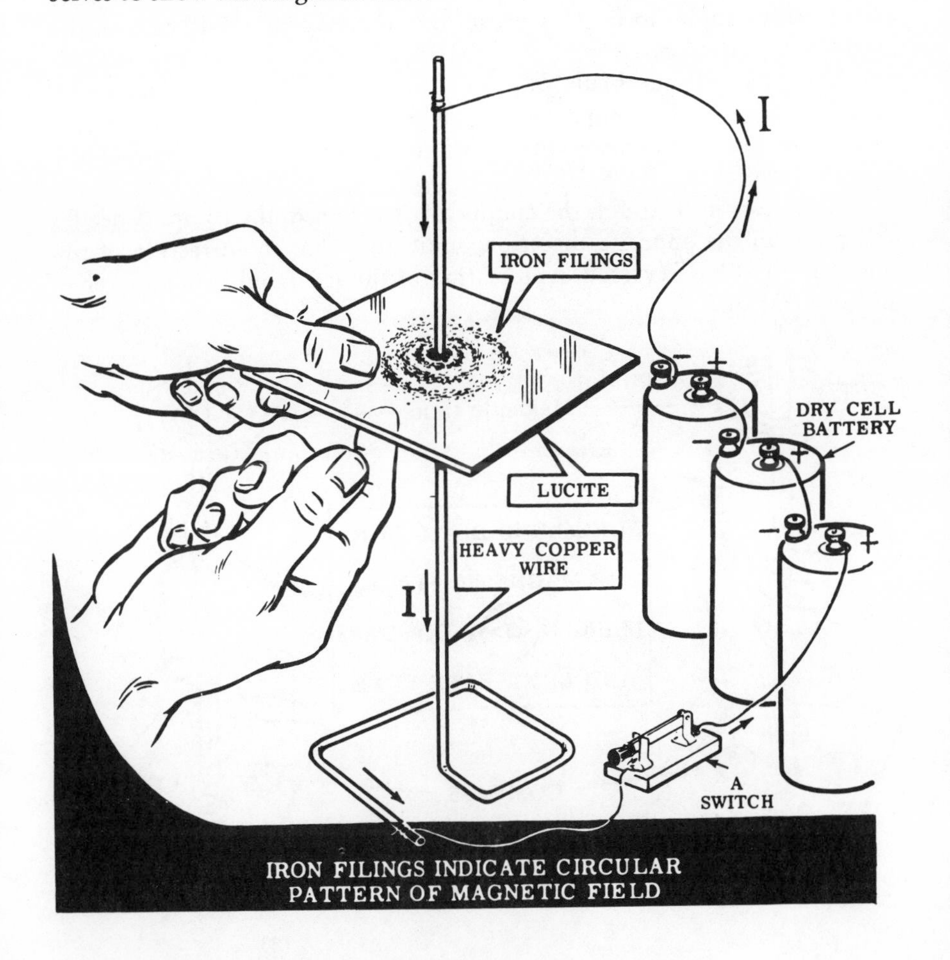

IRON FILINGS INDICATE CIRCULAR PATTERN OF MAGNETIC FIELD

Magnetic Fields around a Conductor (continued)

To demonstrate experimentally the direction of the magnetic field around the current-carrying conductor, a compass needle can be used instead of iron filings.

A compass needle is nothing more than a small bar magnet that will line itself up with the lines of force in a magnetic field. You know from the previous experiment that the magnetic field is circular. Therefore, the compass needle always will be positioned at right angles to the current-carrying conductor.

If you remove the iron filings from the lucite sheet, and the compass is placed on the lucite about 2 inches, or approximately 5 centimeters, away from the conductor, you can trace the direction of the magnetic field around the conductor. With no current flowing, the North pole end of the compass needle will point to the Earth's magnetic North pole. When current flows through the conductor, the compass needle lines itself up at right angles to a radius drawn from the conductor. If the compass needle is moved around the conductor, the needle always maintains itself at right angles to it. This proves that the magnetic field around the conductor is circular.

Using the left-hand rule you can check the direction of the magnetic field indicated by the compass needle. The direction in which the fingers go around the conductor is the same as that of the North pole of the compass needle.

If the current through the conductor is reversed, the compass needle will point in the opposite direction, indicating that the direction of the magnetic field has reversed. Application of the left-hand rule will verify this observation.

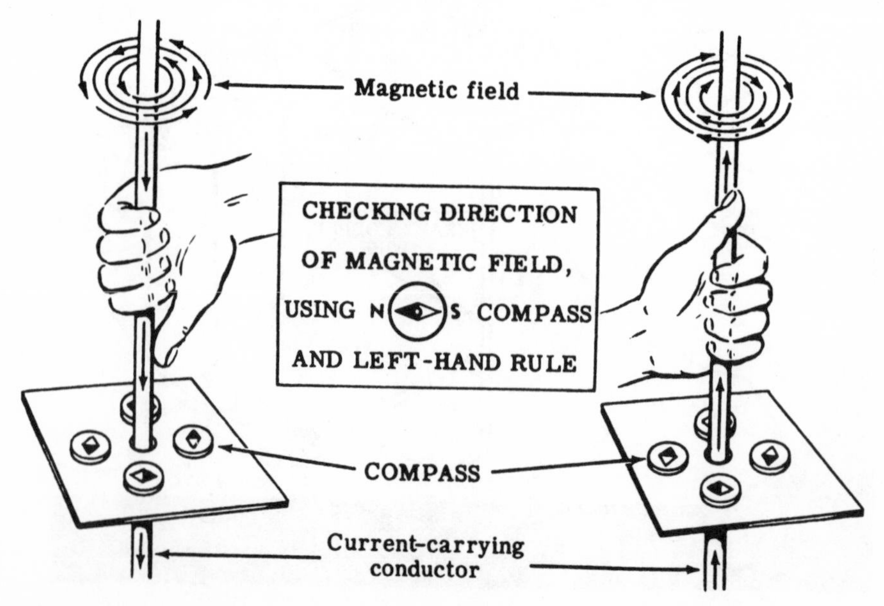

Magnetic Fields around a Coil

Magnetic fields around a coil of wire are extremely important in many pieces of electrical equipment. A coil of wire carrying a current acts as a magnet. If a length of wire carrying a current is bent to form a loop, the lines of force around the conductor all leave at one side of the loop and enter at the other side. Thus, the loop of wire carrying a current will act as a weak magnet having a North pole and a South pole. The North pole is on the side where the lines of force leave the loop and the South pole is on the side where they enter the loop.

If you desire to make the magnetic field of the loop stronger, you can form the wire into a coil of many loops as shown. Now the individual fields of each loop add together and form one strong magnetic field inside and outside the loop. In the spaces between the turns, the lines of force are in opposition and cancel each other out. The coil acts as a strong magnet with the North pole being the end where the lines of force leave the loop.

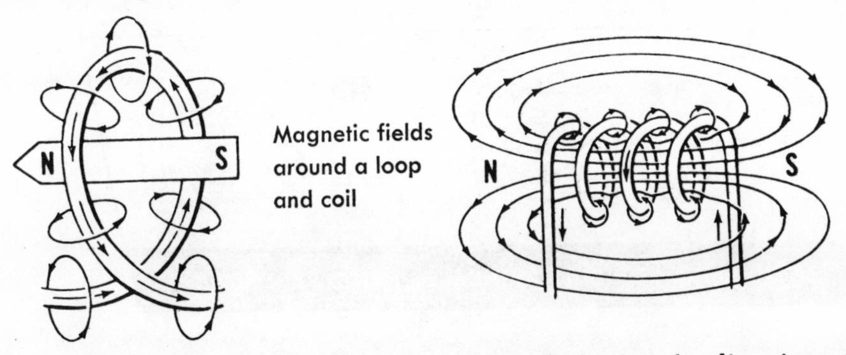

Magnetic fields around a loop and coil

A left-hand rule also exists for coils to determine the direction of the magnetic field. If the fingers of the left hand are wrapped around the coil in the direction of the current flow, the thumb will point toward the North pole end of the coil.

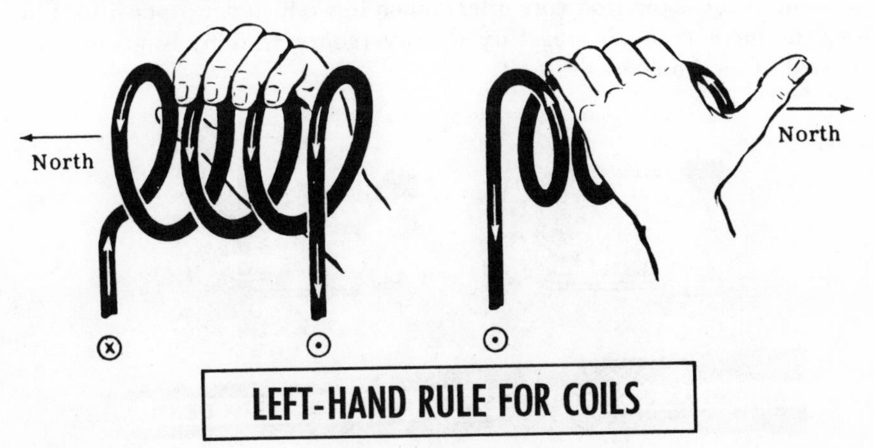

LEFT-HAND RULE FOR COILS

Magnetic Fields around a Coil (continued)

Adding more turns to a current-carrying coil increases the number of lines of force, causing it to act as a stronger magnet. An increase in current also strengthens the magnetic field. Strong electromagnets have coils of many turns and carry as large a current as the wire size permits.

In comparing coils using the same core or similar cores, a unit called the *ampere-turn* is used. This unit is the product of the current in amperes and the number of turns on the wire.

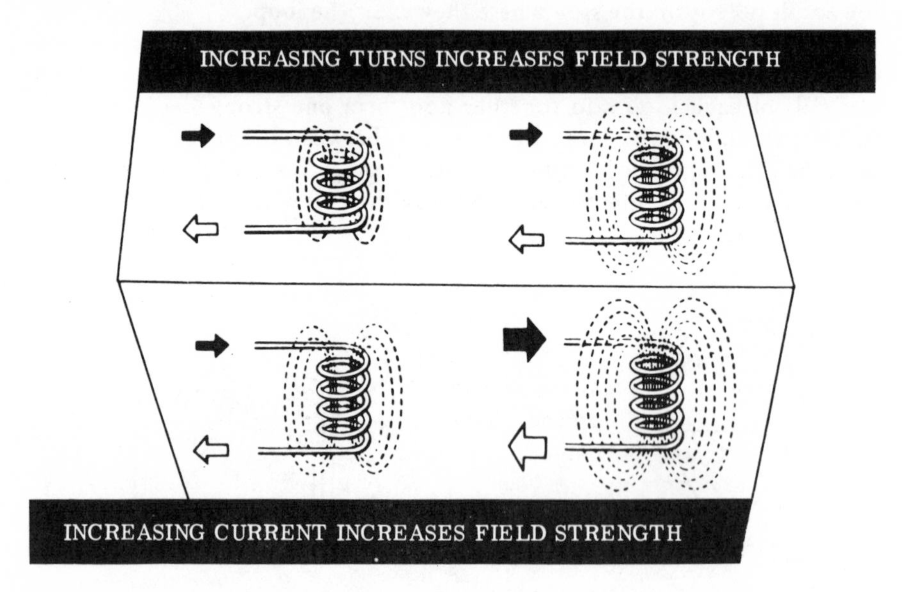

Although the field strength of an electromagnet is increased by using both a large current flow and many turns to form the coil, these factors do not usually concentrate the field enough for use in a practical device. To further increase the flux density, an iron core is inserted in the coil. Because the iron core offers much less reluctance (opposition) to lines of force than air, the flux density (concentration) is greatly increased in the iron core.

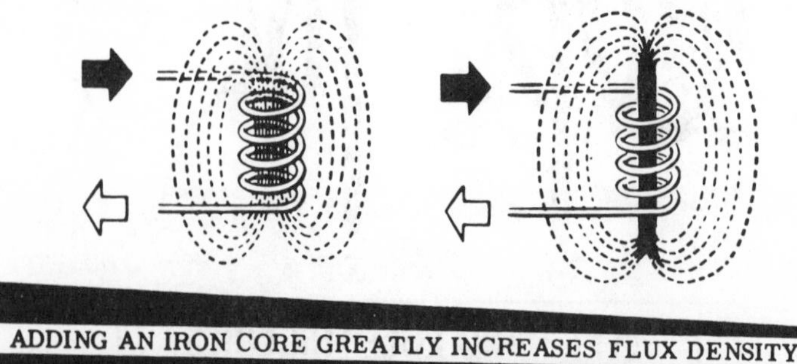

Magnetic Fields around a Coil (continued)

You can show what the field around a coil carrying current is like by taking a piece of wire and forming it into a coil that is threaded through some holes in a piece of lucite plastic.

The rest of the circuit is the same as those showing the fields around a conductor. When iron filings are sprinkled on the lucite and current is passed through the coil, tapping the lucite will cause the iron filings to line up parallel to the lines of force. If you did this, you would observe that the iron filings have formed the same pattern of a magnetic field that exists around a bar magnet.

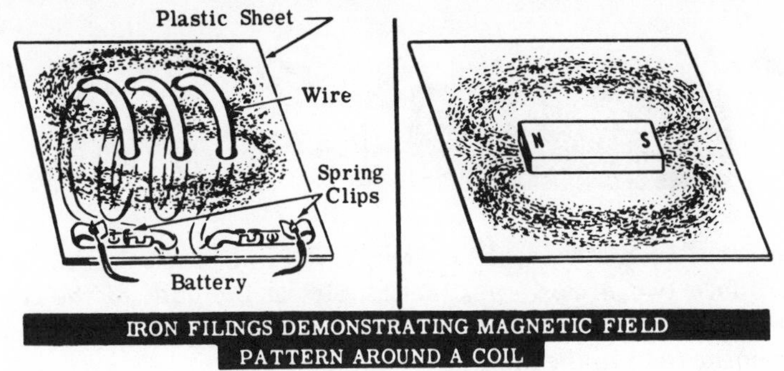

IRON FILINGS DEMONSTRATING MAGNETIC FIELD PATTERN AROUND A COIL

If the filings are removed, and a compass is placed inside the coil, the needle will line up along the axis of the coil with the North pole end of the compass pointing to the North pole end of the coil. Remember that the lines of force inside a magnet or coil flow from the South pole to the North pole. The North pole end of the coil can be verified by using the left-hand rule for coils. If the compass is placed outside the coil and moved from the North pole to the South pole, the compass needle will follow the direction of a line of force as it moves from the North pole to the South pole. When the current through the coil is reversed, the compass needle will also reverse its direction.

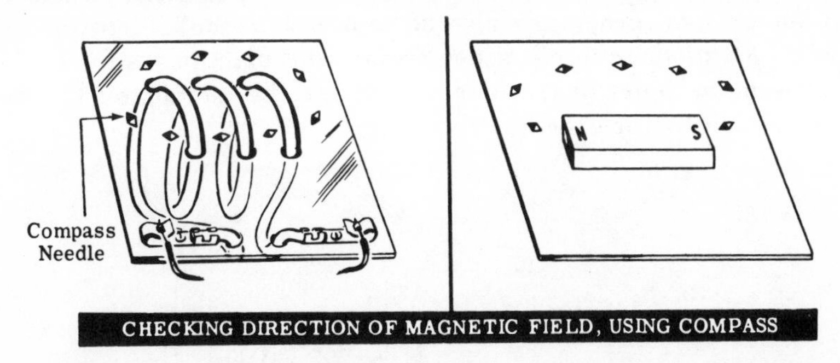

CHECKING DIRECTION OF MAGNETIC FIELD, USING COMPASS

If you placed a soft iron core inside the coil and tested the fields, you would observe that there is a strong concentration of the magnetic field in the iron core, as you would expect from your study of the effects of iron on magnetic fields.

Electromagnets

If the iron core is bent to form a horseshoe and two coils are used, one on each leg of the horseshoe-shaped core as illustrated, the lines of force will travel around the horseshoe and across the air gap, causing a very concentrated field to exist across the air gap. The shorter the air gap, the greater the flux density between the poles.

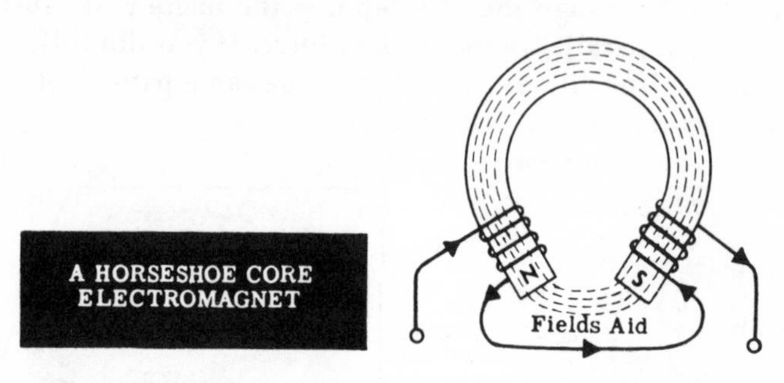

A HORSESHOE CORE ELECTROMAGNET

To cause such a field, the current flow in the series-connected coils must produce two opposite magnetic poles at the ends of the core. Reversing either coil would cause the two fields to oppose each other, canceling out the field in the air gap.

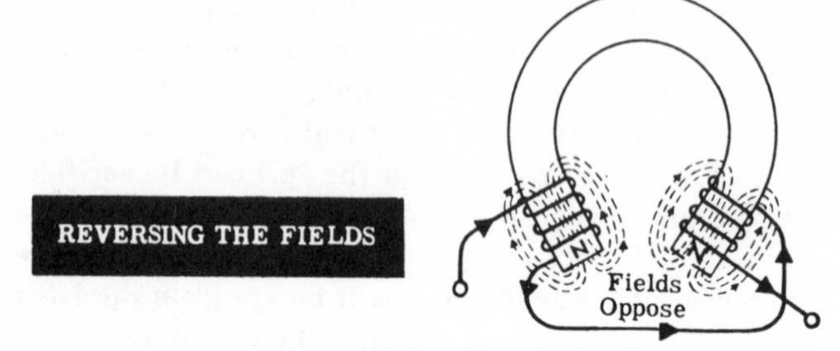

REVERSING THE FIELDS

Electric meters make use of horseshoe-type permanent magnets. Electric motors and generators also make use of a similar type of electromagnet. All of these applications require the placement of a coil of wire between the poles of the magnet and use the interaction between them either to utilize or generate electricity.

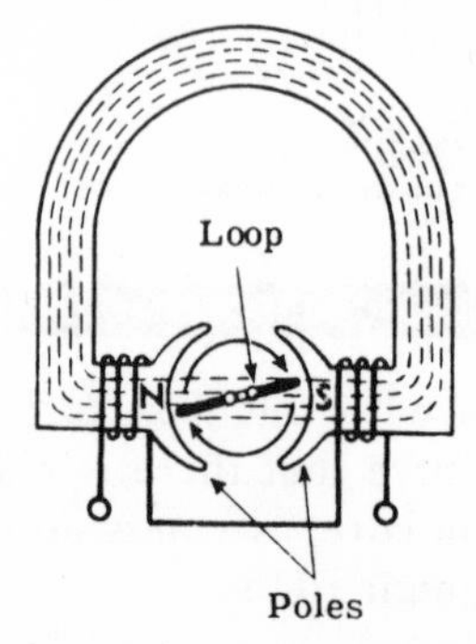

ELECTROMAGNET POLES

Review of Electromagnetism

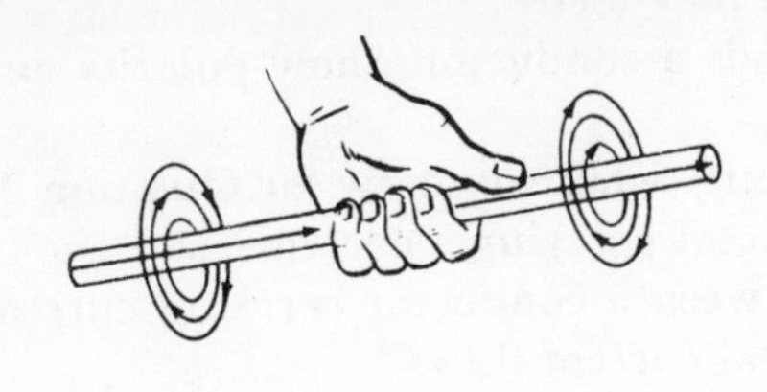

1. ELECTROMAGNETIC FIELD —Current flowing through a wire generates a magnetic field whose direction is determined by the direction of the current flow. The direction of the generated magnetic field is found by using the left-hand rule for a current-carrying conductor.

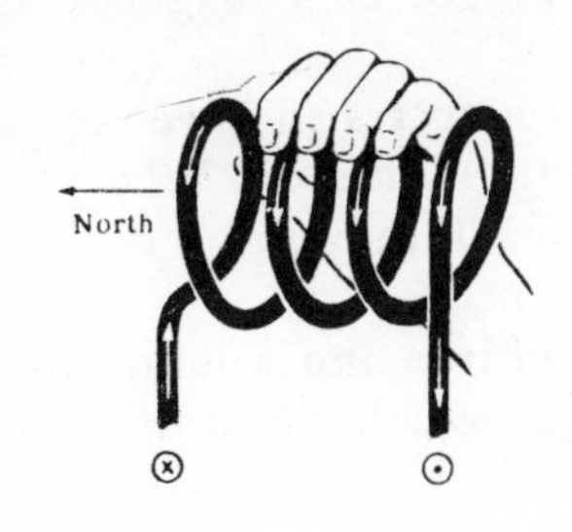

2. MAGNETIC FIELD OF A LOOP OR COIL—A loop generates a magnetic field exactly the same as a bar magnet. If many loops are added in series forming a coil, a stronger magnetic field is generated. The left-hand rule for a coil is used to determine the coil's magnetic polarity.

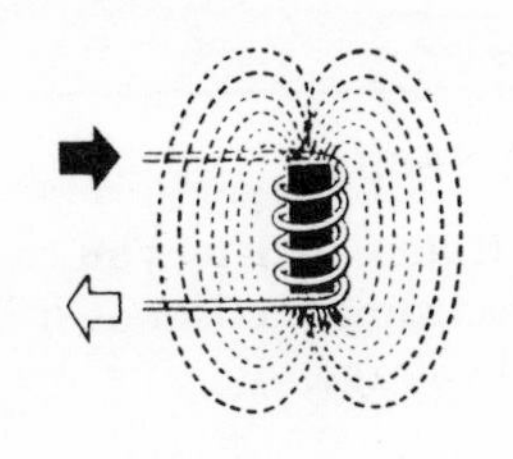

3. FIELD STRENGTH—Increasing the number of turns of a coil increases the field strength and increasing the coil current also increases the field strength. An iron core may be inserted to concentrate the field greatly (increase flux density) at the ends of the coil. The ampere-turn is the unit used in comparing the strength of electromagnetic fields.

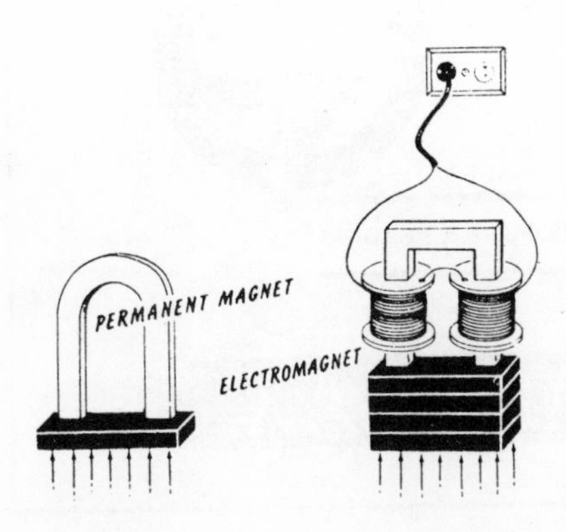

4. PERMANENT-MAGNETS and ELECTROMAGNETIC FIELDS— Electromagnetic fields are much stronger than the permanent magnet type, and are used in most practical electrical machinery. When electromagnets are used, the field strength can be varied by varying the amount of current flow through the field coils.

Self-Test—Review Questions

1. What are electromagnets and why are they used?
2. Draw the magnetic field that surrounds a conductor. Show polarity and direction of current and field.
3. Describe an experiment that would verify what you drew for Question 2.
4. What is the left-hand rule for a conductor carrying a current?
5. Invent a rule for the relationship between a conductor carrying current and its field for the case of conventional current flow.
6. We have stated that the magnetic field around a coil will be greater than for a wire (under the same current conditions). Why?
7. What is the rule for determining the direction of the field around a coil?
8. What are the factors that determine the strength of a magnetic field around a coil carrying current?
9. Would you expect a 2-turn coil carrying 10 amperes to have the same field strength as a coil of 20 turns carrying 1 ampere? Why? What would be the relative field strength of the 2 coils if the current in the second coil was changed to 2 amperes? To 0.5 ampere?
10. If you were to take a horseshoe shaped piece of iron and wound coils on each of the free ends and then connected them together and to a source of electricity, how would you explain the following observations? (a) Current flowing, but no magnetic field or very weak magnetic field. (b) The polarity of the poles of the electromagnet reverses when the current is reversed. (c) The polarity of the poles reverses when the coils are wound in the opposite direction.

Learning Objectives—Next Section

Overview—Now that you know about electromagnetism, you can learn about an important application of electromagnetism as it is used in meters to measure current flow and voltage.

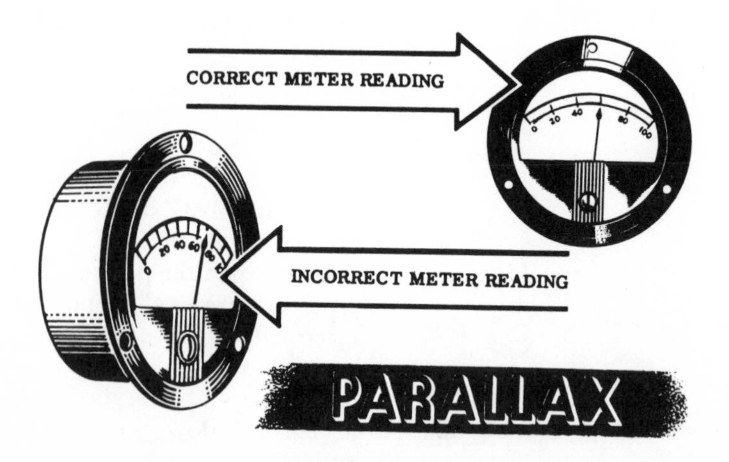

The Basic Meter Movement

You now have learned enough facts about electricity and magnetism to study a very important practical application in meters—the devices that measure current flow.

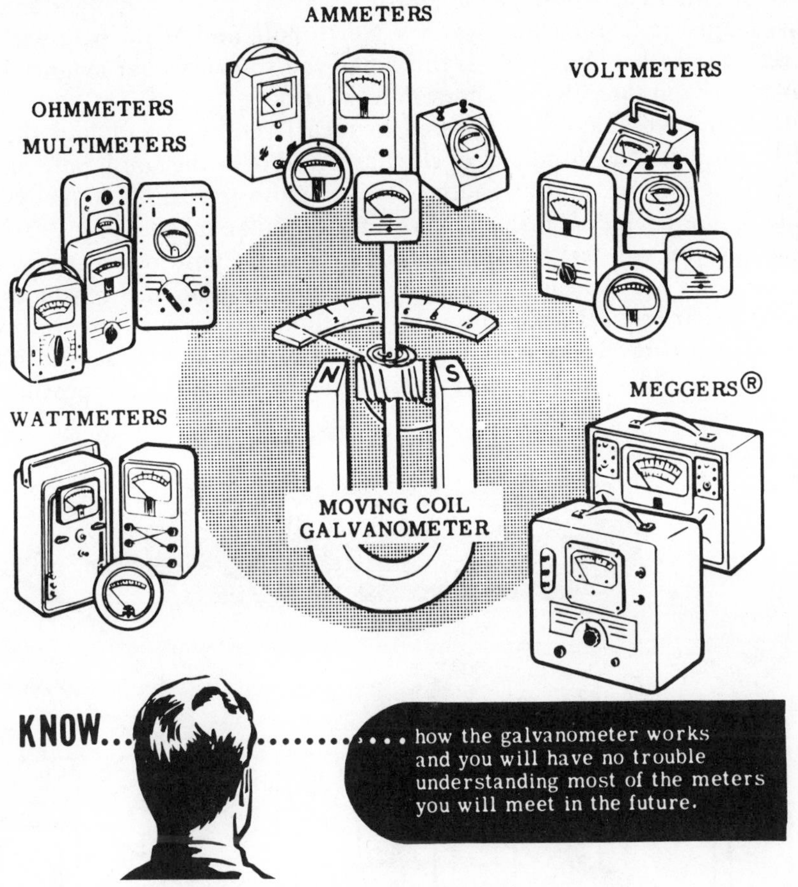

You have probably used meters to show you whether or not an electric current was flowing and how much current was flowing. As you proceed further with your work in electricity, you will find yourself using meters more and more often. Meters are the right hand of anyone working in electricity or electronics, so now is the time for you to find out how they operate.

All the meters you have used and nearly all the meters you will use are made with the same type of meter *works* or movement. This meter movement is based on the principles of an electric current measuring device called the *moving-coil galvanometer*. Nearly all modern meters use the moving-coil galvanometer as a basic meter movement, so once you know how it works, you will have no trouble understanding most of the meters you will be using in the future.

The Basic Meter Movement (continued)

The galvanometer works on the principle of magnetic attraction and repulsion. According to this principle, which you have already learned, like poles repel each other and unlike poles attract each other. This means that two magnetic North poles will repel each other as will two magnetic South poles, while a North pole and South pole will attract one another. You can see this when you suspend a bar magnet on a pivot between the poles of a horseshoe magnet.

If the bar magnet is allowed to turn freely, you will find that it turns until its North pole is as close as possible to the South pole of the horseshoe magnet, and its South pole is as close as possible to the North pole of the horseshoe magnet. If you turn the bar magnet to a different position, you will feel it trying to turn back to the position where the opposite poles are as near as possible to each other. The further you try to turn the bar magnet away from this position, the greater force you will feel. The greatest force will be felt when you turn the bar magnet to the position in which the like poles of each magnet are as close as possible to each other.

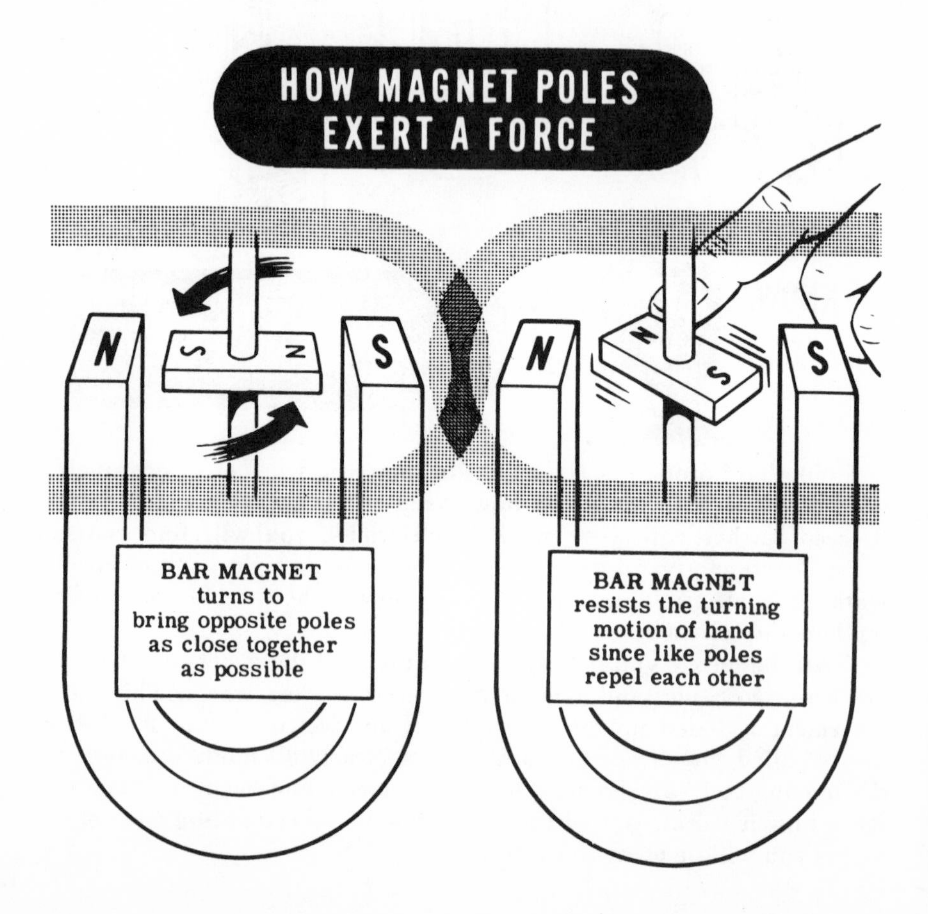

Basic Meter Movement (continued)

The forces of attraction and repulsion between magnetic poles become greater when stronger magnets are used. You can see this when you attach a spring to the bar magnet in such a way that the spring will have no tension when the North poles of the two magnets are as close as possible to each other. With the magnets in this position, the bar magnet would normally turn freely to a position which would bring its North pole as close as possible to the South pole of the horseshoe magnet. With a spring attached, it will turn only part way, to a position where its turning force is balanced by the force of the spring. If you were to replace the bar magnet with a stronger magnet, the force of repulsion between the like poles would be greater and the bar magnet would turn further against the force of the spring.

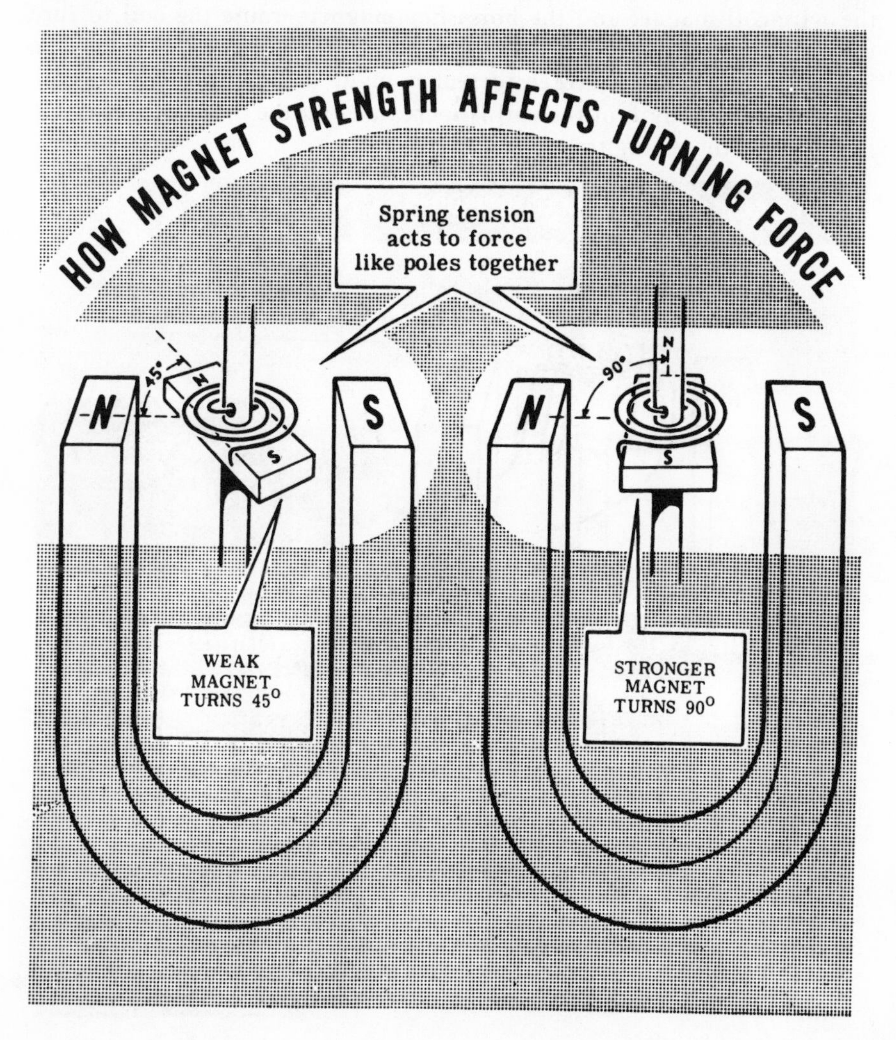

Basic Meter Movement (continued)

If you remove the bar magnet and replace it with a coil of wire, you have a *galvanometer*. Whenever an electric current flows through this coil of wire, it acts as a magnet. The strength of this wire-coil magnet depends on the size, shape, and number of turns in the coil and the amount of electric current flowing through the coil. If the coil itself is not changed in any way, the magnetic strength of the coil will depend on the amount of current flowing through the coil. The greater the current flow in the coil, the greater the magnetic strength of the wire-coil magnet.

If there is no current flow in the coil, it will have no magnetic strength and the coil will turn to a position where there will be no tension on the spring. If you cause a small electric current to flow through the coil, the coil becomes a magnet and the magnetic forces—between the wire-coil magnet and the horseshoe magnet—cause the coil to turn until the magnetic turning force is balanced by the force due to tension in the spring. When a larger current is made to flow through the coil, the magnetic strength of the coil is increased and the wire coil turns further against the spring tension.

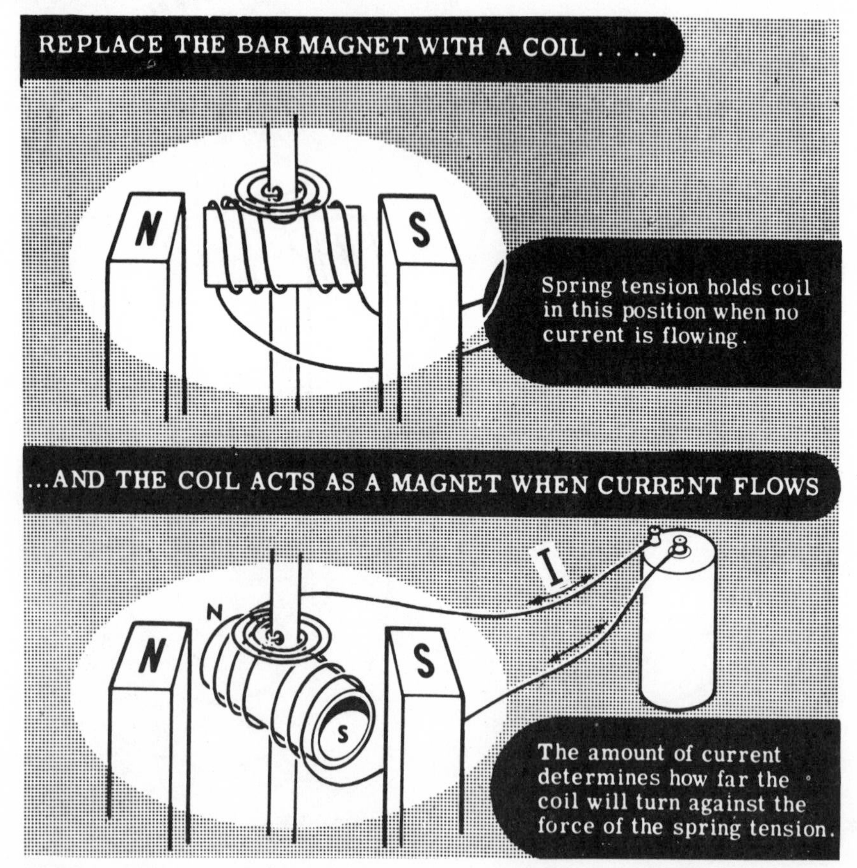

Basic Meter Movement (continued)

When you want to find out how much current is flowing in a circuit, all you need to do is to connect the coil into the circuit and measure the angle through which the coil turns away from its position at rest. It is very difficult to measure this angle, and to calculate the amount of electric current which causes the coil to turn through this angle. However, by connecting a pointer to the coil and adding a scale for the pointer to travel across, you can read the amount of current directly from the scale.

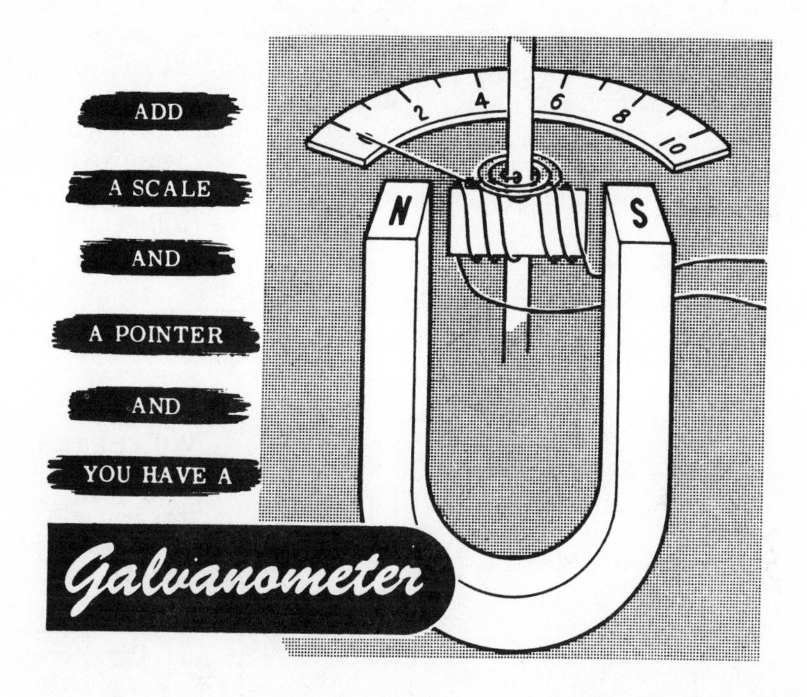

Now that you have added a scale and a pointer, you have a basic dc meter, known as the *D'Arsonval-type* movement, which depends upon the operation of magnets and their magnetic fields. Actually, there are two magnets in this type of meter: one, a stationary permanent horseshoe magnet; the other, an electromagnet. The electromagnet consists of turns of wire wound on a frame, and the frame is mounted on a shaft fitted between two permanently mounted jewel bearings. A lightweight pointer is attached to the coil, and turns with it to indicate the amount of current flow. Current passing through the coil causes it to act as a magnet with poles being attracted and repelled by those of the horseshoe magnet. The strength of the magnetic field about the coil depends upon the amount of current flow. A greater current produces a stronger field, resulting in greater forces of attraction and repulsion between the coil sides and the magnet's poles.

Basic Meter Movement (continued)

The magnetic forces of attraction and repulsion cause the coil to turn so that the unlike poles of the coil and magnet will be brought together. As the coil current increases, the coil becomes a stronger magnet and turns further because of the greater magnetic forces between the coil and magnet poles. Since the amount by which the coil turns depends upon the amount of coil current, the meter indicates the current flow directly.

Although all meters that you will encounter work on the principle of two interacting magnetic fields, there are some modifications that you may encounter. For example, you may ask why the permanent magnet is not associated with the moving part of the meter and the electromagnet on the fixed part. This would eliminate the need to carry current through the springs. Actually there are meters built and used like those described above. These meters, however, are much less sensitive because the strength of the permanent magnet is very limited as a result of its small size; and also, they are less accurate because it is difficult to control the magnetic fields involved. Meters of this type are very inexpensive and are found in less expensive equipment.

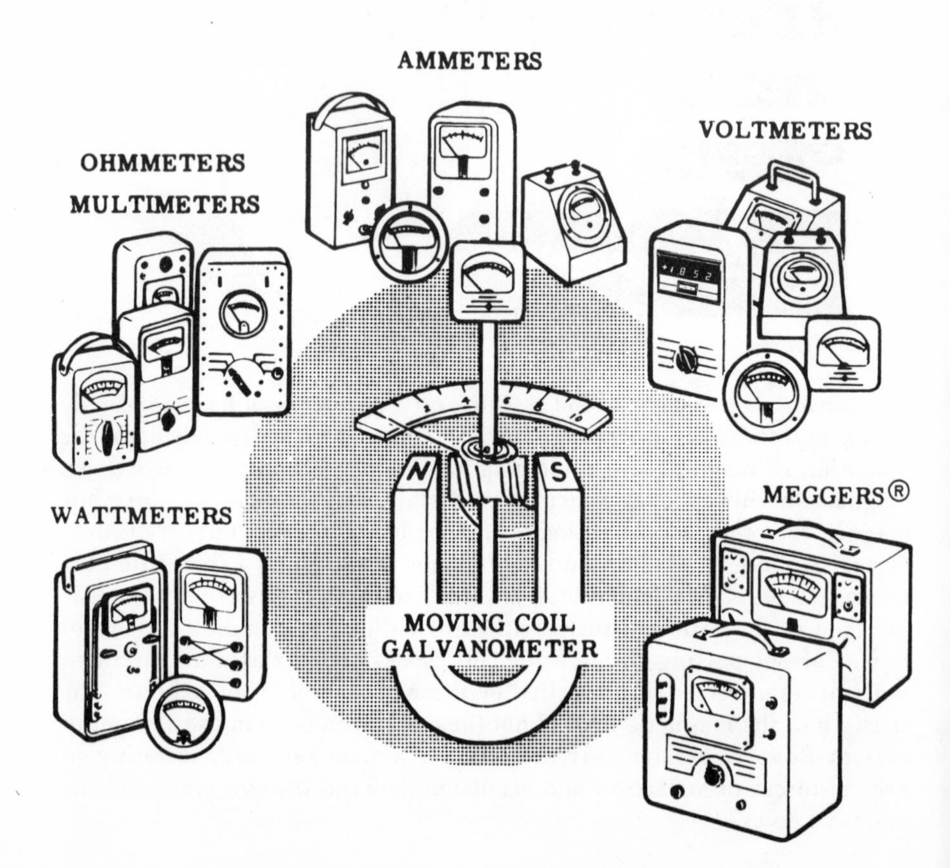

Meter Movement Considerations

While galvanometers are useful in laboratory measurements of extremely small currents, they are not portable, compact, or rugged enough for use in the field. A modern meter movement uses the principles of the galvanometer but is portable, compact, rugged, and easy to read. The coil is mounted on a shaft fitted between two permanently mounted jewel bearings. A lightweight pointer is attached to the coil and turns with the coil to indicate the amount of current flow.

Balance springs on each end of the shaft exert opposite turning forces on the coil and, by adjusting the tension of one spring, the meter pointer may be adjusted to read zero on the meter scale. Since temperature change affects both coil springs equally, the turning effect of the springs on the meter coil is canceled out. As the meter coil turns, one spring tightens to provide a retarding force, while the other spring releases its tension. In addition to providing tension, the springs are used to carry current from the meter terminals through the moving coil.

In order that the turning force will increase uniformly as the current increases, the horseshoe magnet poles are shaped to form semicircles. This brings the coil as near as possible to the North and South poles of the permanent magnet. The amount of current required to turn the meter pointer to full-scale deflection depends upon the magnet strength and the number of turns of wire in the moving coil.

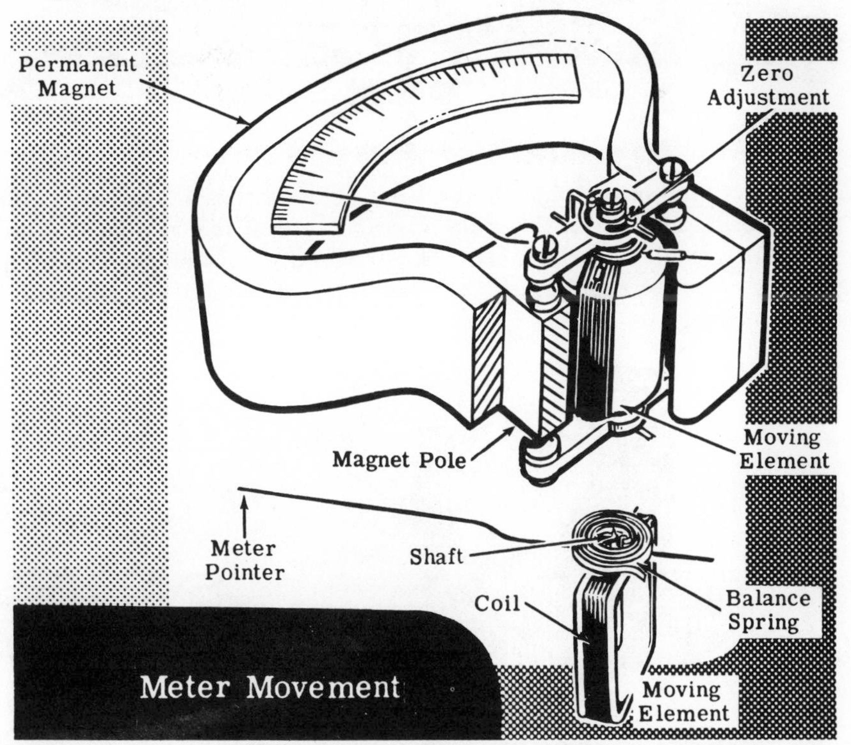

Meter Movement

How Meter Scales Are Read

When you work with electricity, it is necessary that you take accurate meter readings to determine whether equipment is working properly, and to discover what is wrong with equipment that is not operating correctly. Many factors can cause meter readings to be inaccurate. It is necessary to keep them in mind whenever you use a meter. You will find the usable range of a meter scale does not include the extreme ends of the scale. For nearly all meters, the most accurate readings are those taken near the center of the scale. When current is measured with an *ammeter*, *milliammeter*, or *microammeter*, the range of the meter used should be chosen to give a reading near mid-scale.

All meters cannot be used in both horizontal and vertical positions. Due to the mechanical construction of many meters, the accuracy will vary considerably with the position of the meter. Normally, panel-mounted meters are calibrated and adjusted for use in a vertical position. Meters used in many test sets and in some electrical equipment are made for use in a horizontal position.

A zero-set adjustment on the front of the meter is used to set the meter needle at zero on the scale when no current is flowing. This adjustment is made with a small screwdriver and should be checked when using a meter, particularly if the vertical or horizontal position of the meter is changed.

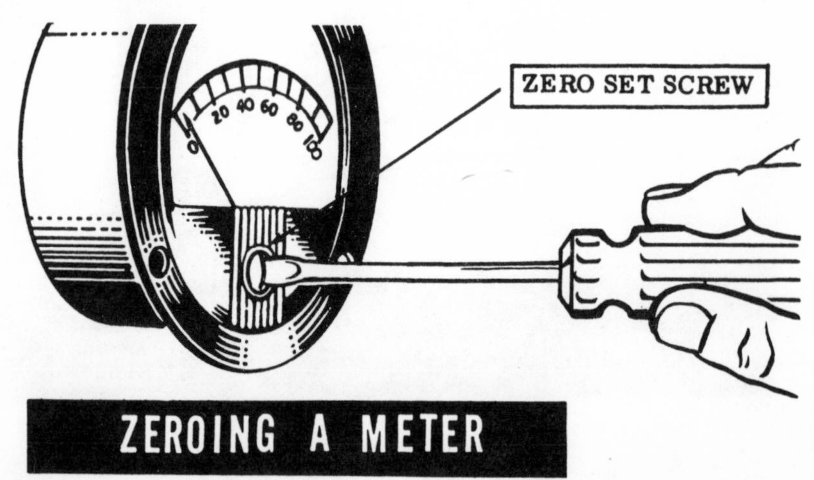

How Meter Scales Are Read (continued)

Meter scales are usually divided into equal divisions, ordinarily with a total of between 30 and 50 divisions. The meter should always be read from a position at right angles to the meter face. Since the meter divisions are small and the meter pointer is raised above the scale, reading the position *from an angle* will result in an inaccurate reading—often as much as an entire scale division. This type of incorrect reading is called *parallax*. Most meters are slightly inaccurate due to the meter construction, and additional error from a parallax reading may result in a very inaccurate reading. Some precision meters have a mirror built into the scale. To read these meters you line up the needle and its image in the mirror with your eye before you read the meter. This eliminates the parallax problem.

When the meter pointer reads a value of current between two divisions of the scale, usually the nearest division is used as the meter reading. However, if a more accurate reading is desired, the position of the pointer between the divisions is estimated, and the deflection between the scale divisions is added to the lower scale division. Estimating the pointer position is called *interpolation*, and you will use this process in many other ways in working with electricity.

Usable Meter Range

The range of an ammeter (an ammeter measures current in amperes) indicates the maximum current which can be measured with the meter. Current in excess of this value can cause serious damage to the meter. If an ammeter has a range of 0-15 amperes, it will measure any current flow which does not exceed 15 amperes; but a current greater than 15 amperes can damage the meter.

While the meter scale may have a range of 0-15 amperes, its useful range for purposes of measurement will be from about 1 ampere to 14 amperes. When this meter scale indicates a current of 15 amperes, the actual current may be much greater but the meter can only indicate to its maximum range. For this reason, the useful maximum range of any meter is slightly less than the maximum range of the meter scale. A current of 0.1 ampere on this meter scale would be very difficult to read since it would not cause the meter needle to move far enough from zero to obtain a definite reading.

Smaller currents such as 0.001 ampere would not cause the meter needle to move and, thus, could not be measured at all with this meter. The useful minimum range of a meter never extends down to zero, but extends, instead, only to the point at which the reading can be readily distinguished from zero.

Review of How a Meter Works

Although you may have little reason to open up a meter to repair it, it is very important that you understand the principle of operation so that you can know how properly to use and take care of meters. As you might suspect, meters are delicate and must be handled with care. Meters are used everywhere electricity is used or generated to tell us what is going on in electrical systems. Therefore, it is imperative that you thoroughly understand how to read meters properly.

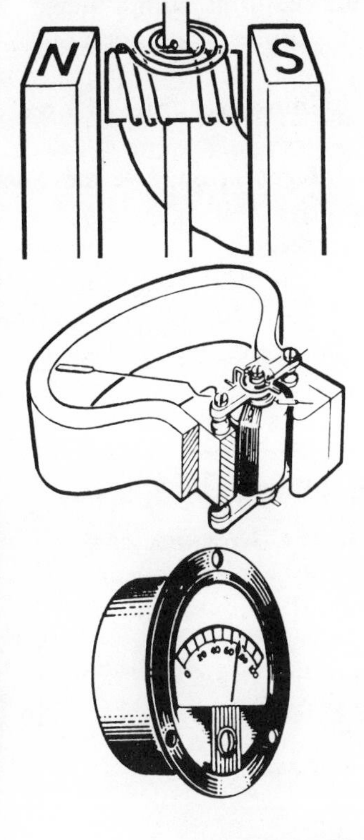

1. METER COIL—Moving coil which acts as a magnet when current flows in the coil.

2. METER MOVEMENT — Current-measuring instrument consisting of a moving coil suspended between the poles of a horseshoe magnet. Current in the coil causes the coil to turn.

3. PARALLAX — Meter reading error due to taking a reading from an angle.

4. INTERPOLATION—Estimating the meter reading between two scale divisions.

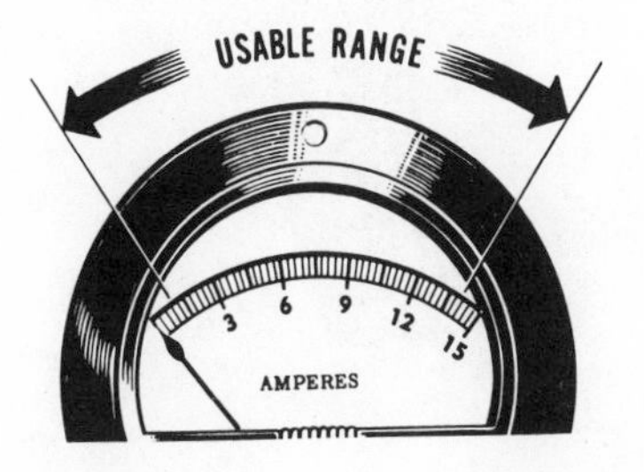

5. USEFUL METER RANGE

Self-Test—Review Questions

1. Why are meters important electrical devices?
2. What is the principle of operation of the moving-coil meter?
3. Why do you think it is necessary to handle meters with care?
4. What is the result of putting a current through a meter that is slightly in excess of the maximum scale reading? What is the result if the current is much greater?
5. Can meters be used in any position? Explain.
6. What should you do if you are ready to take some readings with a meter and you find that with no current flowing, the pointer does not read zero?
7. What is parallax? What is the effect of parallax on meter readings? How do you avoid it?
8. What is interpolation of meter readings? Does interpolation give more accurate results? Is interpolation always necessary?
9. What is normally considered the useful range of a meter?
10. In the meters we have been considering, what do you think would happen if you reversed the leads going to the meter in a circuit that had given a normal reading earlier? Would you say that a meter would have to be hooked up in a specific way?

Learning Objectives—Next Section

Overview—You have learned how a meter works. Now you can learn how it is connected and used to measure current flow. A meter connected to measure current flow is called an *ammeter*.

THE AMMETER

Measuring the Units of Current Flow

You have learned that the unit of current flow is the ampere and that this corresponds to the movement of charge (electrons) at the rate of 1 coulomb per second and that 1 coulomb is equal to 6.28 million, million, million electrons. You have also learned how current flow produces magnetic fields and how this principle is used in the construction of the meter. We will now examine how these ideas are put to practical use for the measurement of current.

The device which is used to measure the rate of current flow through a conducting material, and to display this information in such a way that you can use it, is called an *ammeter*. An ammeter indicates, in amperes, the number of electrons passing a given point in the material (which is, in practice, almost always a wire).

To be able to do this, the ammeter must somehow be connected into the wire in such a way that it is able to count *all* the electrons passing without letting any of them slip past uncounted. The only way to do this is to break the wire, or *open the line* as it is called, and to insert the ammeter physically in it.

When an ammeter is inserted in this way into a wire being used to carry current to an electric lamp, the ammeter is said to be *in series* with the lamp. The schematic symbol for the ammeter is a circle with the letter "A" or "I" in it.

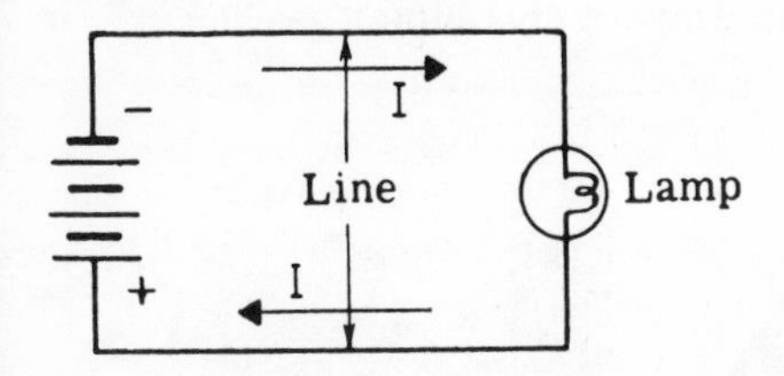

Without ammeter

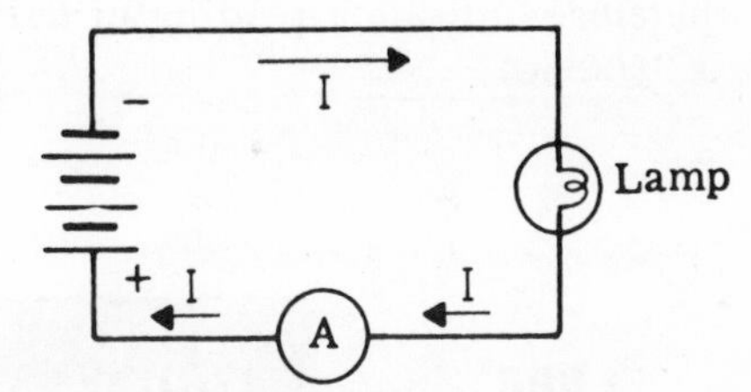

Ammeter connected in series with line to measure lamp current.

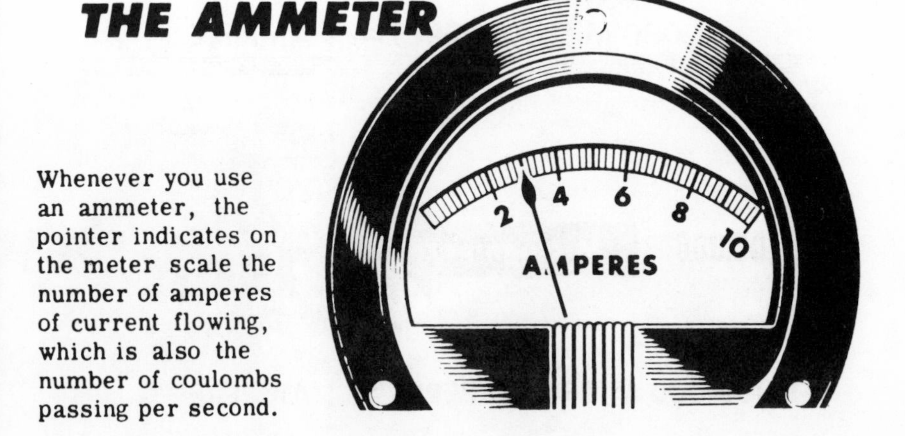

Whenever you use an ammeter, the pointer indicates on the meter scale the number of amperes of current flowing, which is also the number of coulombs passing per second.

How Small Currents Are Measured

While the ampere is the basic unit of measurement for current flow, it is not always a convenient unit to use. Current flows seldom exceed 1,000 amperes but may often be as little as 1/1,000 of an ampere. For measuring currents of less than 1 ampere, some other unit is needed. A cup of water is not measured in gallons, nor is the flow of water from a fire hydrant measured in cups. In any kind of measurement, a *usable* unit of measurement is needed. Since current flow seldom exceeds 1,000 amperes, the ampere can be used satisfactorily as the unit for currents in excess of 1 ampere. However, it is not convenient as the unit for currents of *less* than 1 ampere.

If the current flow is between 1/1,000 of an ampere and 1 ampere, the unit of measure used is the *milliampere* (abbreviated mA), which is equal to 1/1,000 ampere. For current flow of less than 1/1,000 ampere, the unit used is the *microampere* (abbreviated μA), which is equal to 1/1,000,000 ampere. Meters used for measuring milliamperes of current are called *milliammeters*, while meters used for measuring microamperes of current are called *microammeters*. Units of measurement are subdivided in such a way that a quantity expressed in one unit may be readily changed to another unit, either larger or smaller.

Fractions such as halves, quarters, thirds, etc., are seldom used in electrical work; decimals being generally preferred. A meter would therefore indicate a reading of half-an-ampere (½) either as "0.5 A" or as "500 mA."

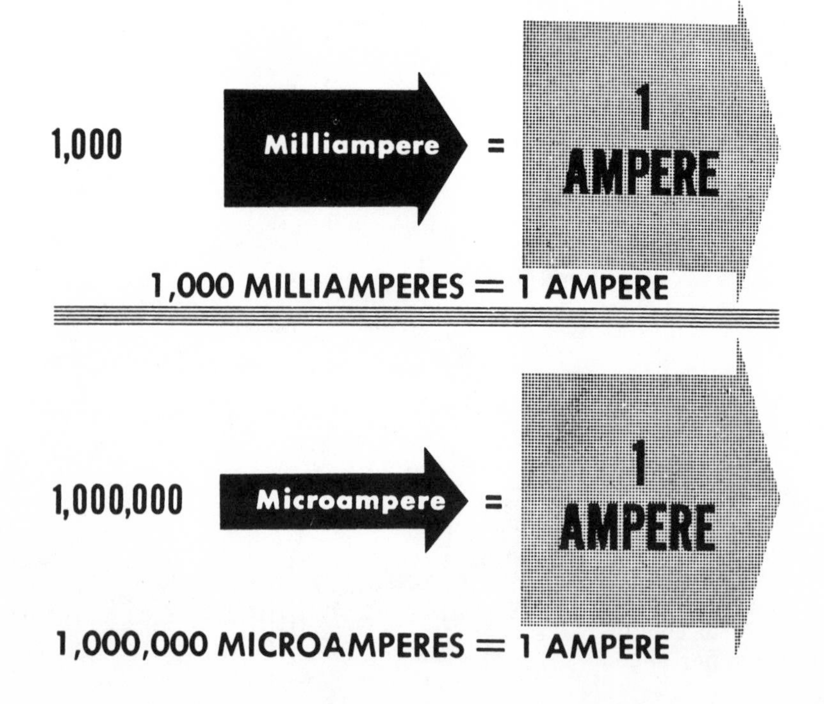

How Units of Current Are Converted

In order to work with electricity, you must be able to convert from one unit of current to another. Since a milliampere (mA) is 1/1,000 of an ampere, milliamperes can be converted to amperes by moving the decimal point three places to the *left*. For example, 35 milliamperes is equal to 0.035 ampere. There are two steps required in order to arrive at the correct answer. First, the original position of the decimal point must be located. The decimal is then moved three places to the left, thereby converting the unit from milliamperes to amperes. If no decimal point is given with the number, it is always understood to follow the last number in the quantity. In the example given, the reference decimal point is after the number 5, and to convert from milliamperes to amperes, it must be moved three places to the left. Since there are only two whole numbers to the left of the decimal point, a zero must be added to the left of the number to provide for a third place, as shown.

CONVERTING MILLIAMPERES TO AMPERES

35 milliamperes = ? ampere

Move decimal point three places to the left.

35. MILLIAMPERES = .035 AMPERE

REFERENCE POINT

When converting amperes to milliamperes, you move the decimal point three places to the *right* instead of the left. For example, 0.125 ampere equals 125 milliamperes, and 16 amperes equals 16,000 milliamperes. In these examples, the decimal point is moved three places to the right of its reference position, with three zeros added in the second example to provide the necessary decimal places.

CONVERTING AMPERES TO MILLIAMPERES

.125 ampere = ? milliamperes

Move decimal point three places to the right.

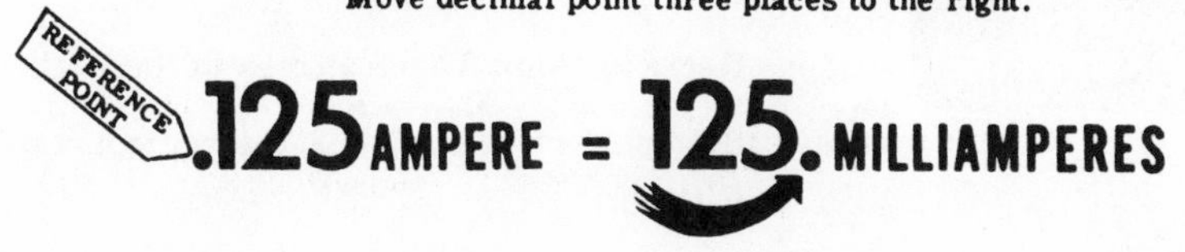

How Units of Current Are Converted (continued)

Suppose that you are working with a current of 125 microamperes and you need to express this current in amperes. If you are converting from a larger unit to a smaller unit, the decimal point is moved to the *right*; while to change from a smaller unit to a larger unit, the decimal point is moved to the *left*. Since a microampere is 1/1,000,000 ampere, the ampere is the larger unit. Then converting microamperes to amperes is a conversion from small to large units and the decimal point should be moved to the left. In order to convert millionths to units, the decimal point must be moved six decimal places to the left; thus 125 microamperes equals 0.000125 ampere. The reference point in 125 microamperes is after the 5, and in order to move the decimal point six places to the left, you must add three zeros ahead of the number 125. When converting microamperes to milliamperes, the decimal point is moved only three places to the left; thus 125 microamperes equals 0.125 milliampere.

If your original current is in amperes and you want to express it in microamperes, the decimal point should be moved six places to the right. For example, 3 amperes equals 3,000,000 microamperes, because the reference decimal point after the 3 is moved six places to the right with the six zeros added to provide the necessary places. To convert milliamperes to microamperes, the decimal point should be moved three places to the right. For example, 125 milliamperes equals 125,000 microamperes, with the three zeros added to provide the necessary decimal places.

CONVERTING UNITS OF CURRENT

MICROAMPERES TO AMPERES

Move Decimal Point Six Places to the Left.

125. microamperes = **.000125** ampere

MICROAMPERES TO MILLIAMPERES

Move Decimal Point Three Places to the Left.

125. microamperes = **.125** milliampere

AMPERES TO MICROAMPERES

Move Decimal Point Six Places to the Right.

3. amperes = **3,000,000.** microamperes

MILLIAMPERES TO MICROAMPERES

Move Decimal Point Three Places to the Right.

125. milliamperes = **125,000.** microamperes

Milliammeters and Microammeters

An ammeter having a meter scale range of 0-1 ampere is actually a milliammeter with a range of 0-1,000 milliamperes. As fractions are seldom used in electricity, a meter reading of ½ ampere on the 0-1 ampere range is given as 0.5 ampere or 500 milliamperes (mA). For ranges less than 1 ampere, milliammeters and microammeters are used to measure current.

If you are using currents between 1 milliampere and 1,000 milliamperes, milliammeters are used to measure the amount of current. For currents of less than 1 milliampere, microammeters of the correct range are used. Very small currents of 1 microampere or less are measured on special laboratory-type instruments called *galvanometers*. You will not normally use the galvanometer, since the currents used in electrical equipment are usually between 1 microampere and 100 amperes and, thus, can be measured with a microammeter, milliammeter, or ammeter of the correct range. Meter scale ranges for milliammeters and microammeters, like ammeters, are in multiples of 5 or 10 since these multiples are easily converted to other units.

In using a meter to measure current, the maximum reading of the meter range should always be higher than the maximum current to be measured. A safe method of current measurement is to start with a meter having a range *much greater* than you expect to measure, in order to determine the correct meter to use.

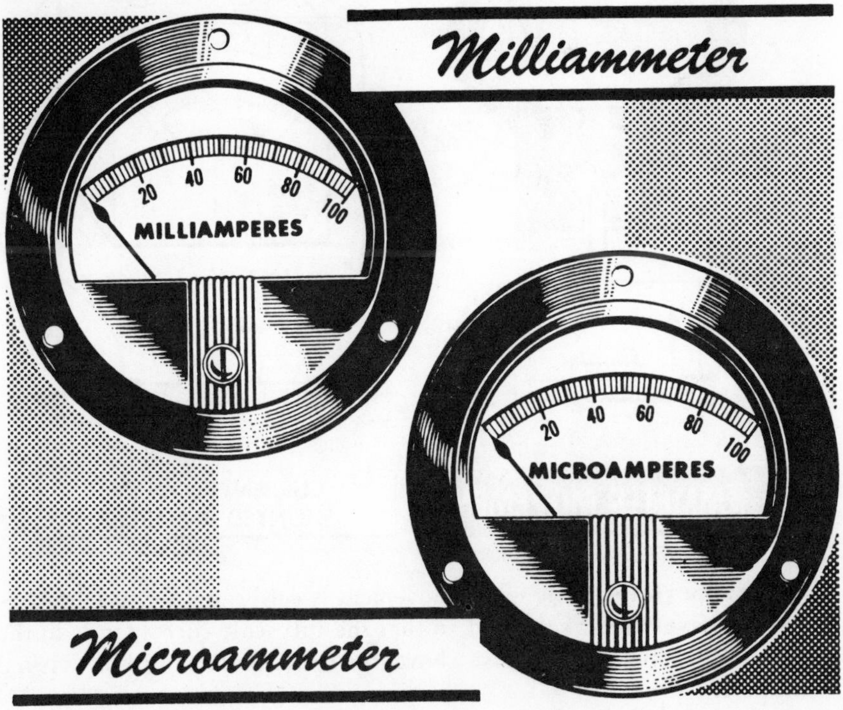

Milliammeter

20 40 60 80 100
MILLIAMPERES

20 40 60 80 100
MICROAMPERES

Microammeter

How Ammeter Ranges Are Converted

Meter ranges could be converted by using magnets of different strength or by changing the number of turns in the coil, since either of these changes would alter the amount of current needed for full-scale deflection. However, the wire used in the coil must always be large enough to carry the maximum current of the range the meter is intended for; therefore, changing the wire size would only be practical in the small current ranges, since large wire cannot be used as a moving coil. To keep the wire size and the coil small, basic meter movements are normally limited to a range of 1 milliampere or less. Also, for using a meter for more than one range, it is impractical to change the magnet or the coil each time the range is changed.

For measuring large currents, a low range meter is used with a *shunt*, which is a heavy wire connected across the meter terminals to carry most of the current. This shunt allows only a small part of the current to actually flow through the meter coil. Usually a 0-1 milliampere meter is used, with the proper-sized shunt connected across its terminals to achieve the desired range. The 0-1 milliammeter is a basic meter movement which you will find in many types of meters you will use. Other common basic current ranges are 0-100 μA and 0-50 μA.

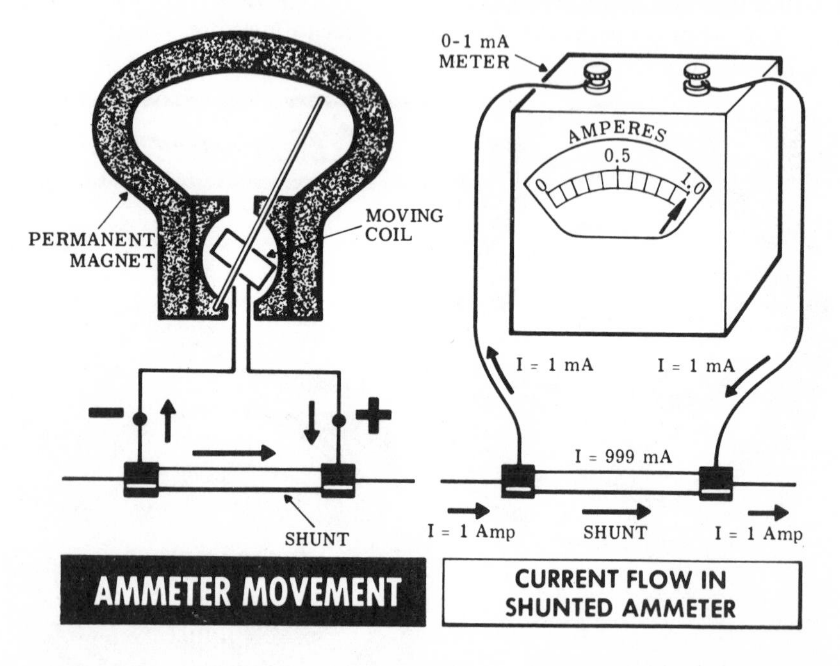

AMMETER MOVEMENT

CURRENT FLOW IN SHUNTED AMMETER

Even though the basic meter movement is calibrated for 0-1 mA, it is usual to have the dial marked so that the full scale corresponds to the value with the shunt. In the case above, the meter scale would be marked 0-1 ampere.

Multirange Ammeters

You have seen that you can change the range of an ammeter by the use of shunts. The range will vary according to the value of the shunt. Some ammeters are built with a number of internal shunts and a switching arrangement that is used to parallel different shunts across the meter movement to measure different currents. Thus, a single meter movement can be used as a *multirange* ammeter. A scale for each range is painted on the meter face. The diagram below shows a multirange ammeter with a 0-3, 0-30, 0-300 ampere range. Note the three scales on the meter face.

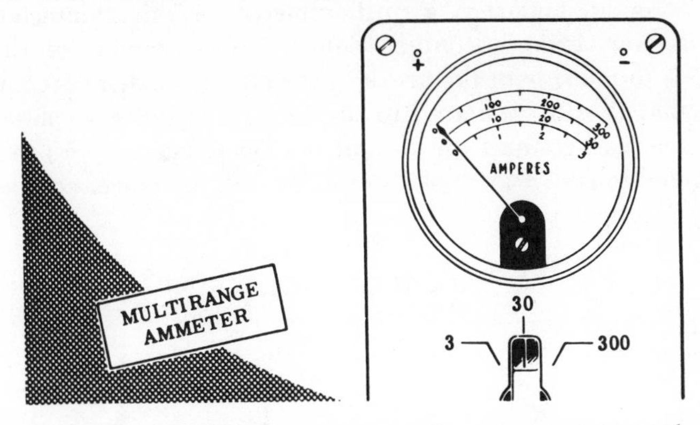

When a multirange ammeter is used to measure an unknown current, the *highest* range is always used first, then the next highest range, and so on, until the needle is positioned about midscale. In this way you can be assured that the current is not excessive for the meter range, and you will never have the unfortunate experience of burning out a meter movement, or of wrapping the needle around the stop-peg.

Some multimeters use external shunts and do away with internal shunts and the switching arrangement. Changing range for such a meter involves shunting it with the appropriate shunt. In the diagram, the ammeter is calibrated to read 30 amperes full-scale by shunting it with the 30-ampere shunt.

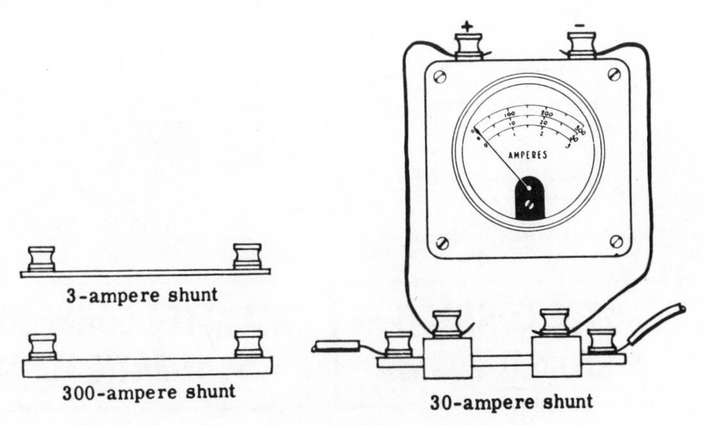

How Ammeters Are Connected into Circuits

All of the ammeters that have been described are called *direct-current meters*; that is, they are designed for circuits where the direction of the current flow is *constant*. When you see a meter of this type, you will notice that the meter terminals are marked with (+) and (—). These markings tell you how the meter should be connected into the electric circuit. You will remember that for the meter to read properly, it is necessary that the magnetic fields have the proper polarity. This means that the current must pass through the meter in the correct direction. The rule for connecting an ammeter, a milliammeter, or microammeter in the proper direction is very simple. Connect the terminal of the meter marked (—) to the side of the break in the circuit made to accommodate the meter that is still connected to the (—) or negative terminal of the power source; and connect the side of the meter marked (+) to the side still connected to the (+) or positive terminal of the power source.

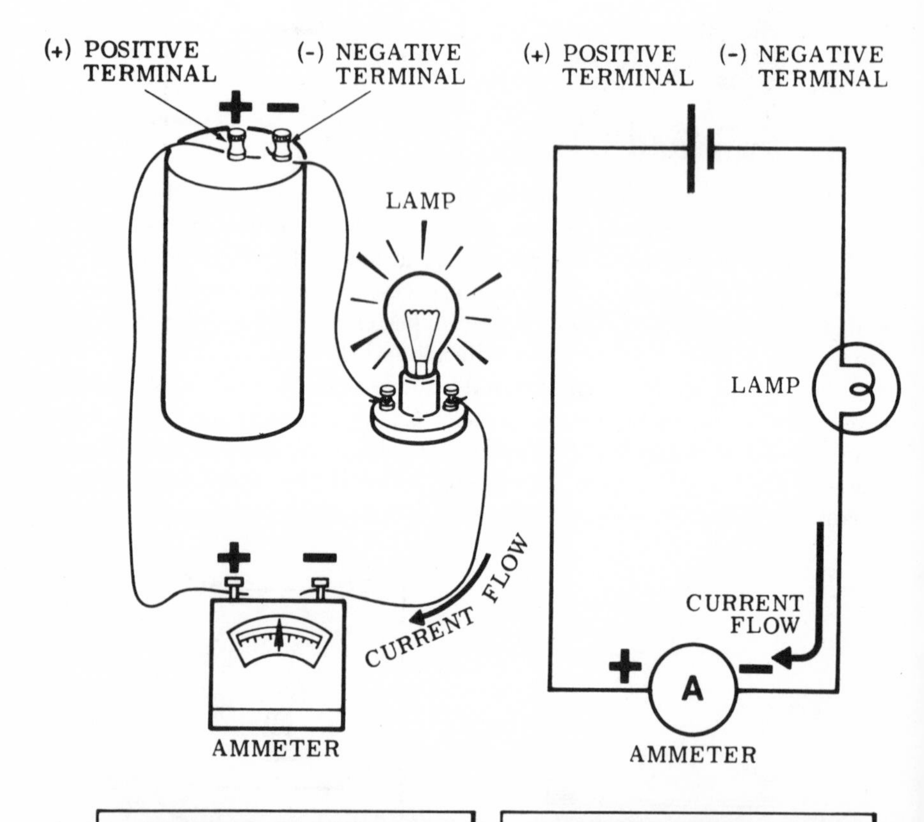

AMMETER CONNECTION PICTORIAL FORM

AMMETER CONNECTION SCHEMATIC FORM

Review of How Current Is Measured

To review what you have discovered about how current is measured, consider some of the important facts you have studied.

1. AMPERE—Unit of rate of flow of electrons, equal to 1 coulomb per second.

$$1 \text{ ma} = \frac{1}{1000} \text{ amp}$$

2. MILLIAMPERE—A unit of current equal to 1/1,000 ampere.

$$1 \ \mu\text{a} = \frac{1}{1{,}000{,}000} \text{ amp}$$

3. MICROAMPERE—A unit of current equal to 1/1,000,000 ampere.

4. AMMETER—A meter used to measure currents of 1 ampere and greater.

5. MILLIAMMETER—A meter used to measure currents between 1/1,000 ampere and 1 ampere.

6. MICROAMMETER—A meter used to measure currents between 1/1,000,000 ampere and 1/1,000 ampere.

Review of How Current is Measured (continued)

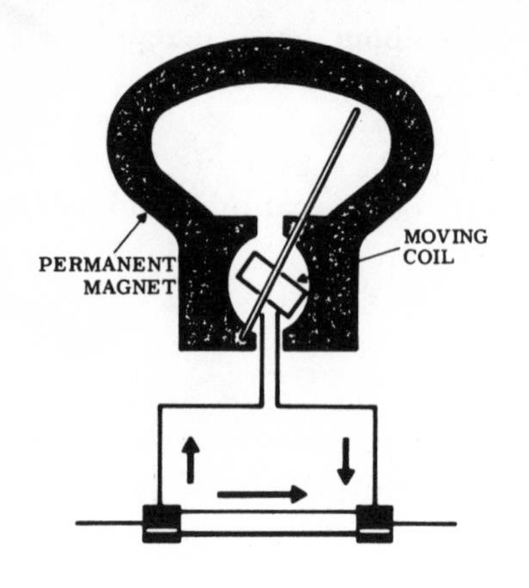

7. BASIC AMMETER MOVEMENT—0-1 mA (milliammeter) movement with shunt wire across the meter terminals to increase the meter scale range.

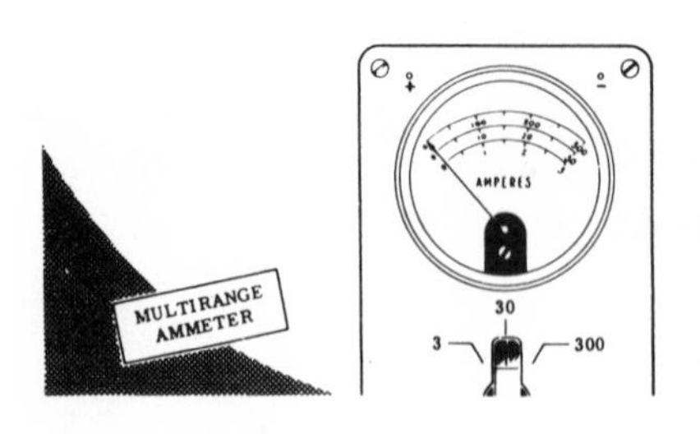

8. MULTIRANGE AMMETER—A single meter movement used for measuring different current ranges. Each range requires a different shunt. The shunts may be inside the meter movement and controlled by a switching arrangement, or they may be external, in which case they are connected in parallel with the meter binding posts.

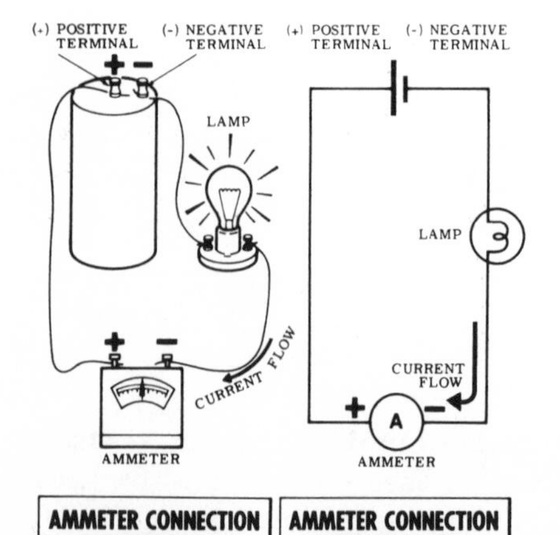

9. AMMETER CONNECTION—Ammeters are always connected in series with the circuit so that all of the circuit current flows through the ammeter. The connections are made by breaking the circuit and connecting the (—) terminal of the meter to the wire still connected to the (—) terminal of the power source; and similarly, for the (+) terminal of the ammeter.

Self-Test—Review Questions

1. Define the basic unit of current and name the instrument used to measure it.
2. Define the milliampere and the microampere. Why and when are these terms used?
3. Calculate the following conversions:

Convert to amperes	Convert to milliamperes	Convert to microamperes
10,000 mA	0.250 A	0.35 mA
10,000 μA	0.525 A	0.022 mA
2,500 mA	1.330 A	1.000 mA
1,000 mA	0.002 A	13.435 mA
33,500 mA	0.055 A	1.000 A
	1,000 μA	0.035 A
	13,200 μA	

4. Draw an electric circuit showing the way that an ammeter is connected into the system. Use a battery and a lamp as the other circuit elements. Indicate the direction of current flow and the proper polarity of the meter.
5. What are microammeters and milliammeters? How do they differ from an ammeter?
6. Why is a shunt used?
7. Show the way that a shunt is connected to make an ammeter.
8. What are the usual ranges of basic meter movement?
9. Assume that you have a basic meter movement of 0-250 μA, and it is desired to make this into a 0-10 ampere meter. Show how the shunt is connected and how the currents would distribute between the shunt and the meter movement with a current flow of 5 amperes.
10. How do multirange meters differ from single range meters? If you are not sure of the approximate amount of current flowing in the circuit, which range of a multirange meter should you start with if you are going to measure the current in the circuit?

Learning Objectives—Next Section

Overview—Just as an ammeter measures current, a voltmeter measures potential difference or voltage. In the next section, you will find out how a meter is used to measure voltage.

Units of Voltage

The electromotive force between two unequal charges is usually expressed in volts; but, when the difference in potential is only a fraction of a volt, or is more than 1,000 volts, other units are used. For voltages of less than 1 volt, *millivolts* and *microvolts* are used, just as milliamperes and microamperes are used to express currents less than 1 ampere. While current seldom exceeds 1,000 amperes, voltage often exceeds 1,000 volts, so that the kilovolt (abbreviated kV)—equal to 1,000 volts—is used as the unit of measurement. When the potential difference between two charges is between 1/1,000 volt and 1 volt, the unit of measure is the *millivolt* (abbreviated mV). When it is between 1/1,000,000 volt and 1/1,000 volt, the unit is the *microvolt* (abbreviated μV).

Meters for measuring voltage have scale ranges in microvolts, millivolts, volts, and kilovolts, depending on the units of voltage to be measured. Ordinarily, you will work with voltages between 1 and 500 volts and use the volt as a unit. Voltages of less than 1 volt, and more than 500 volts, are not used except in special applications of electrical and electronic equipment.

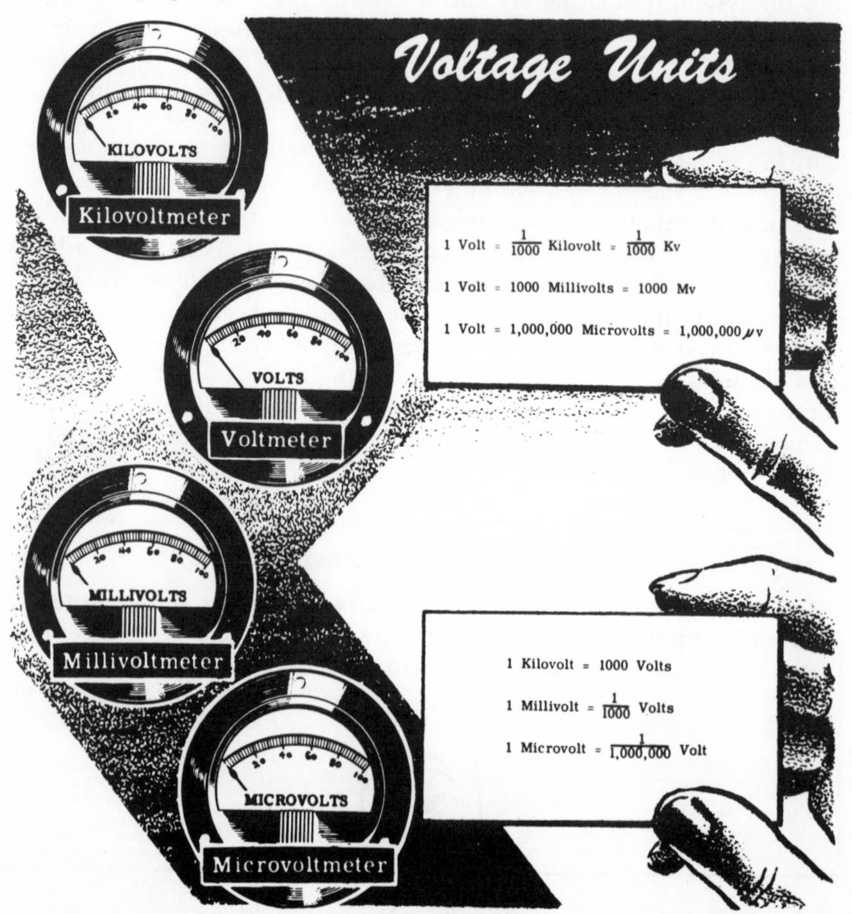

Converting Units of Voltage

Units of voltage measurement are converted in the same way that units of current are converted. In order to convert millivolts to volts, the decimal point is moved three places to the *left*; and to convert volts to millivolts, the decimal point is moved three places to the *right*. Similarly, in converting microvolts to volts, the decimal point is moved six places to the left; and in converting volts to microvolts, the decimal point is moved six places to the right. These examples show that in converting units, the same rules of moving the decimal point apply to *both* voltage and current.

Kilo (meaning one thousand) is not used to express current, but since it is used to express voltage, you may need to know how to convert *kilovolts* to volts, and the reverse. To convert kilovolts to volts, the decimal point is moved three places to the right; and to convert volts to kilovolts, it is moved three places to the left. For example, 5 kilovolts equals 5,000 volts. Since the decimal point is after the 5, three zeros are added to provide the necessary places. Also, 450 volts equals 0.45 kilovolt as the decimal point is moved three places to the left.

CONVERTING VOLTAGE UNITS

VOLTS TO KILOVOLTS

Move the decimal point
3 places to the left.

450 volts = .45 kilovolt

KILOVOLTS TO VOLTS

Move the decimal point
3 places to the right.

5 kilovolts = 5000 volts

VOLTS TO MILLIVOLTS

Move the decimal point
3 places to the right.

15 volts = 15,000 millivolts

MILLIVOLTS TO VOLTS

Move the decimal point
3 places to the left.

500 millivolts =.5 volt

VOLTS TO MICROVOLTS

Move the decimal point
6 places to the right.

15 volts = 15,000,000 microvolts

MICROVOLTS TO VOLTS

Move the decimal point
6 places to the left.

3505 microvolts =.003505 volt

How a Voltmeter Works

As you know, an ammeter measures the rate at which charges move through a material; and that the rate of current flow through a given material varies directly with the voltage difference. That is, the greater the voltage difference, the greater the current flow. Voltage is measured by a meter that is called a *voltmeter.* The voltmeter consists of an ammeter in series with a special piece of material called a *resistor* that limits the current flow. The measurement of voltage is made by measuring the current that flows in the meter circuit. When used in a voltmeter, the resistor (which you will learn about a little later) is called a *multiplier resistor.* For a given ammeter and multiplier resistor, a large current will flow when the voltage is high; a small current will flow when the voltage is low. The meter scale can be marked or calibrated in volts and read directly. Since it is desirable to keep the current in the voltmeter circuit as low as possible so that connection of the voltmeter will not disturb other measurements, the meter used is always a milliammeter or a microammeter.

The multiplier resistor determines the scale range of a voltmeter. Since the multiplier is built into most of the voltmeters you will use, you can measure voltage by making very simple connections. Whenever the (+) meter terminal is connected to the (+) terminal of the voltage source, and the (—) meter terminal to the (—) terminal of the voltage source, with nothing else connected in series, the meter reads voltage directly. When using a voltmeter, it is important to observe the correct meter polarity and to use a meter with a maximum scale range *greater* than the maximum voltage you expect to read.

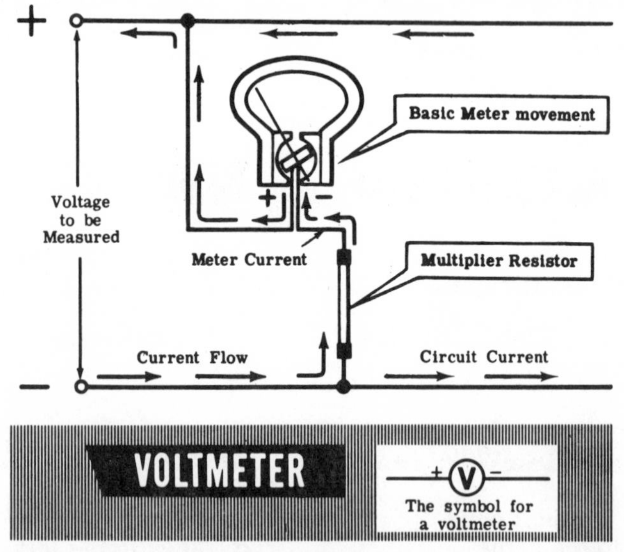

A voltmeter is always connected *across* the voltage source to be measured. When connected this way, the voltmeter is in a *parallel* circuit.

How a Voltmeter Is Used

A voltmeter is used to measure voltage anywhere in a circuit. If it is to measure a source of voltage, such as a battery, the negative (—) side of the voltmeter is always connected to the negative (—) side of the battery; the positive (+) side of the voltmeter is always connected to the positive (+) side of the battery. If connections are *reversed*, meter needle will move to the left of zero mark, and a reading will not be obtained.

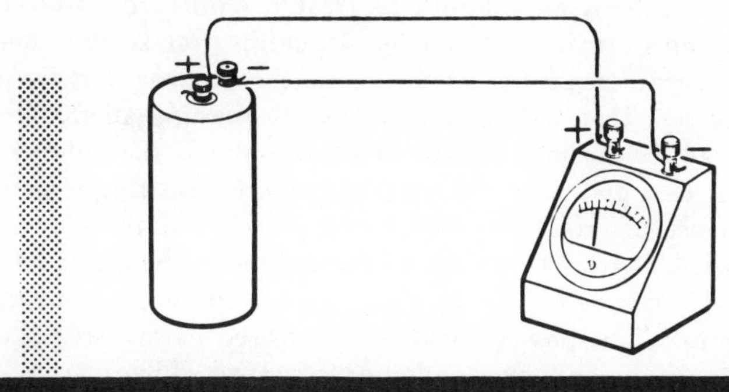

CONNECT A VOLTMETER PLUS TO PLUS - MINUS TO MINUS

The electric circuit through which the current flows is usually called the *load* and may consist of a single item such as a lamp, or may be extremely complex as in the case of all of the devices that are connected to the output of a large generator.

When the voltmeter is to measure the voltage drop across a load, the negative (—) lead is connected to the side of the load where the electrons *enter* (the (—) side); and the positive (+) lead is connected to the side of the load from which the electrons *emerge* (the (+) side).

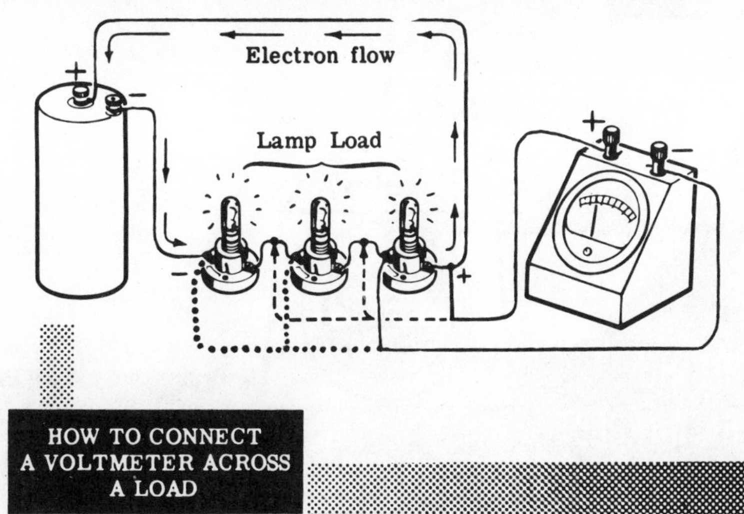

HOW TO CONNECT A VOLTMETER ACROSS A LOAD

Voltmeter Ranges

Just as you learned with the ammeter, it is necessary with a voltmeter to connect it into the circuit with the *proper polarity*; and to choose a voltmeter with the *proper range* for the measurement that is to be made. With direct current voltmeters such as those already discussed, it is a good idea to choose a voltmeter that reads between 10% and 90% of the scale. For example, a voltmeter with a full scale reading of 100 volts would be used to measure voltages between 10 and 90 volts. All meters are delicate instruments and should be treated with care. Improper use of a voltmeter can burn it out or change its calibration so that it will no longer give accurate readings. If the voltage is approximately known, choose a voltmeter that will give a mid-scale reading so that there is plenty of leeway if you have estimated incorrectly. If the voltage is not known, then it is a good idea to start with the *highest* range meter that you have and use progressively lower range meters until a satisfactory range is obtained. It is always a good idea to look at the meter when you throw the switch to see that it is properly connected and that the range is correct. Obviously, the power should be removed as fast as possible if you notice that the meter is overloaded or is connected backwards.

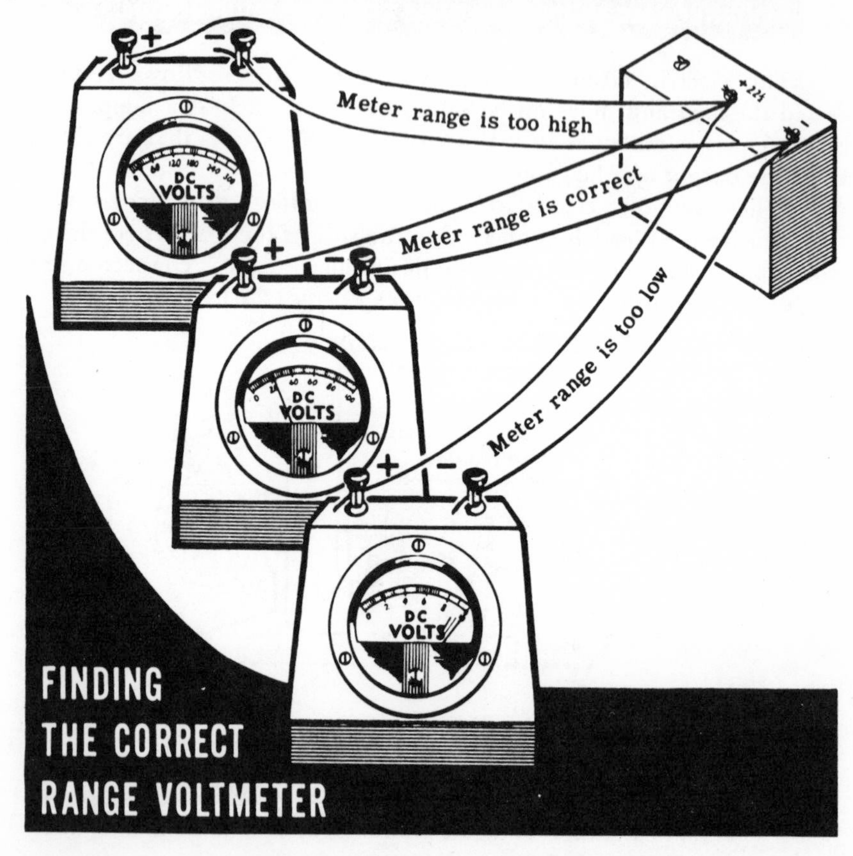

FINDING THE CORRECT RANGE VOLTMETER

Multirange Voltmeters

As described, the range of any voltmeter can be increased by the addition of a multiplier to the voltmeter circuit in series with the basic meter movement. The multiplier causes reduction of the deflection of the pointer on the meter; by using multipliers of known values, the deflection can be reduced as much as desired.

Multirange voltmeters, like multirange ammeters, are instruments which you will frequently use. They are physically very similar to ammeters, and their multipliers are usually located inside the meter with suitable switches, or sets of terminals, on the outside for selecting range. Proper range is selected by starting with the *highest* range and working downward, until the needle reads about midscale.

Because they are lightweight, portable, and can be set up for different voltage ranges by the flick of a switch, multirange voltmeters are extremely useful.

The simplified drawing below shows a three-range, multirange voltmeter.

TYPICAL 3-RANGE MULTIRANGE VOLTMETER

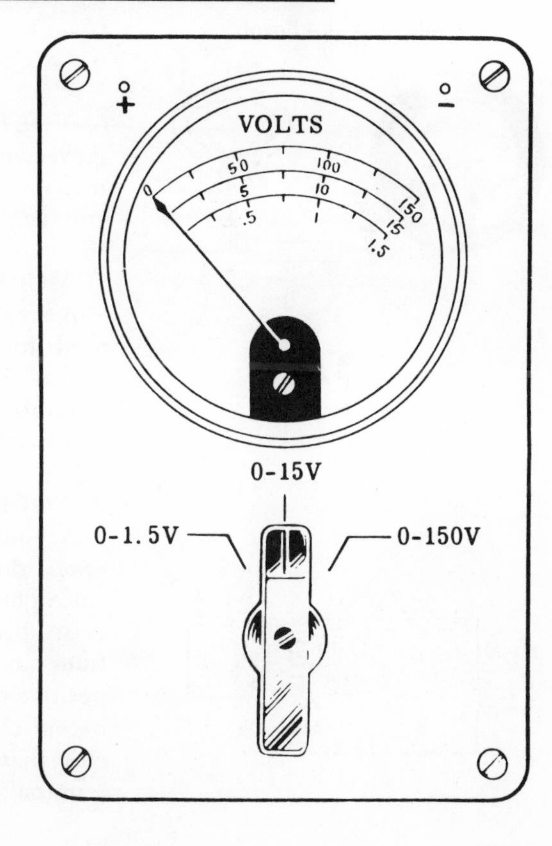

Review of Voltage Units and Measurement

V = Voltage (EMF)

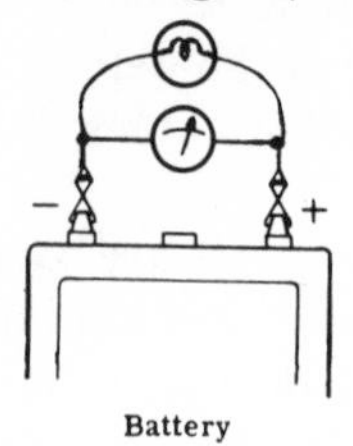

1. VOLT—The unit of potential difference. It is equal to work of 1 joule per coulomb.

2. MAINTENANCE OF EMF—EMF is maintained by having a source of energy that is converted into potential difference to keep the emf constant, regardless of load.

1 millivolt (1 mV) = $\frac{1}{1000}$ volt
1 volt = 1000 millivolts

3. MILLIVOLT—A unit of voltage equal to 1/1,000 volt.

1 microvolt (1 μV) = $\frac{1}{1,000,000}$ volt
1 volt = 1,000,000 microvolts

4. MICROVOLT—A unit of voltage equal to 1/1,000,000 volt.

1 kilovolt (1 kV) = 1000 volts
1 volt = $\frac{1}{1000}$ kilovolt

5. KILOVOLT—A unit of voltage equal to 1,000 volts.

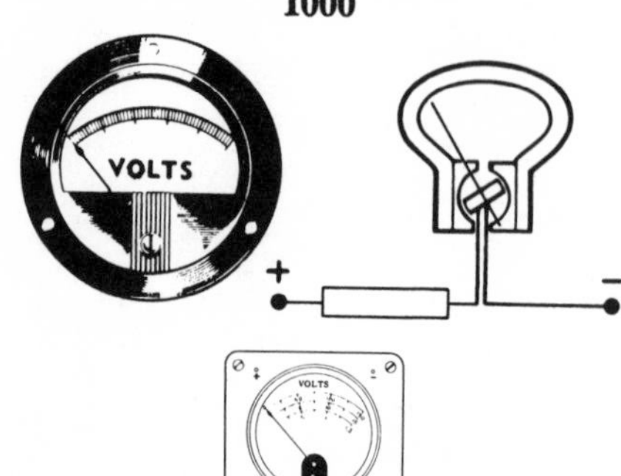

6. VOLTMETER — Basic meter movement with a series-connected multiplier, calibrated to measure voltage.

7. MULTIRANGE VOLTMETER—A single meter movement that is used to measure different voltage ranges. Each range uses a different multiplier that is selected by means of a switch.

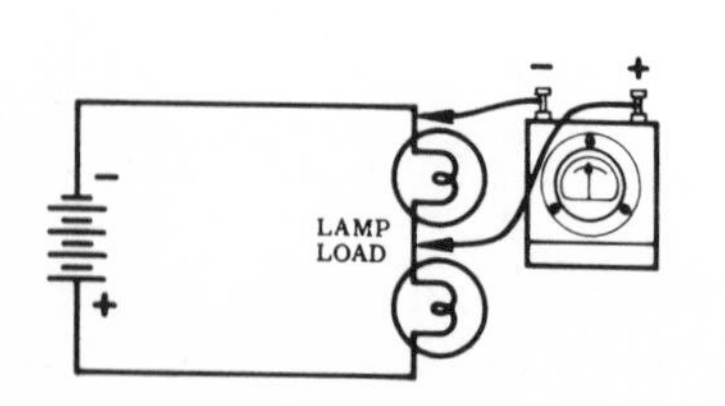

8. VOLTMETER CONNECTION—A voltmeter is always connected *across* the circuit to be measured since potential difference or voltage exists between two points. Connections are always made so that the positive (+) side of the meter is connected closest to the part of the circuit that goes to the positive (+) terminal of the power source.

Self-Test—Review Questions

1. What is the unit of potential difference? How is it defined?
2. What symbols are used to designate voltage?
3. Assume that you have three terminals available (A, B, and C) and that you are measuring the following potential differences with voltmeters of suitable range. What are the potential differences between the end (A to C) in each case? The polarity indications in parenthesis ( ) indicate the way the meter was connected to give a proper upscale reading.

   (a) A(−) to B(+) = 45 volts; B(−) to C(+) = 45 volts
   (b) A(−) to B(+) = 115 volts; B(−) to C(+) = 23.5 volts
   (c) A(+) to B(−) = 7.5 volts; B(−) to C(+) = 20 volts
   (d) A(−) to B(+) = 6 volts; B(+) to C(−) = 6 volts

4. What is the essential factor that allows a battery, or a generator to maintain an emf, while a pair of charged bars cannot?
5. Describe the components that make up a voltmeter and tell how each component functions.
6. Define the following and give the appropriate abbreviation:
   (a) millivolt (b) microvolt (c) kilovolt
7. Calculate the following conversions:

Convert to volts	Convert to millivolts
0.01 kV	0.01 V
500 mV	1 V
25,000 μV	10 V
1,500 mV	10 μV
0.001 kV	1,000 μV
10 kV	320 μV
10 mV	3,200 μV

Convert to kilovolts	Convert to microvolts
100 V	10 mV
17,500 V	0.001 mV
1,500,300 mV	1,450 mV
1,350 V	1 V
100,000 V	0.001 V
1 V	3.25 mV

8. What is the symbol for the voltmeter?
9. How is a voltmeter connected into a circuit? How does this differ from the connection for an ammeter?
10. How would you go about selecting the proper voltmeter to measure an unknown voltage? What sort of meter range would you select if you knew the voltage was approximately 80 volts?

Learning Objectives—Next Section

Overview—The flow of current is controlled by resistance. Devices called *resistors* can be used to do this. They are of basic importance in the study of electricity. In fact almost all of Volume 2 is devoted to the study of resistances and how they are used in electrical circuits.

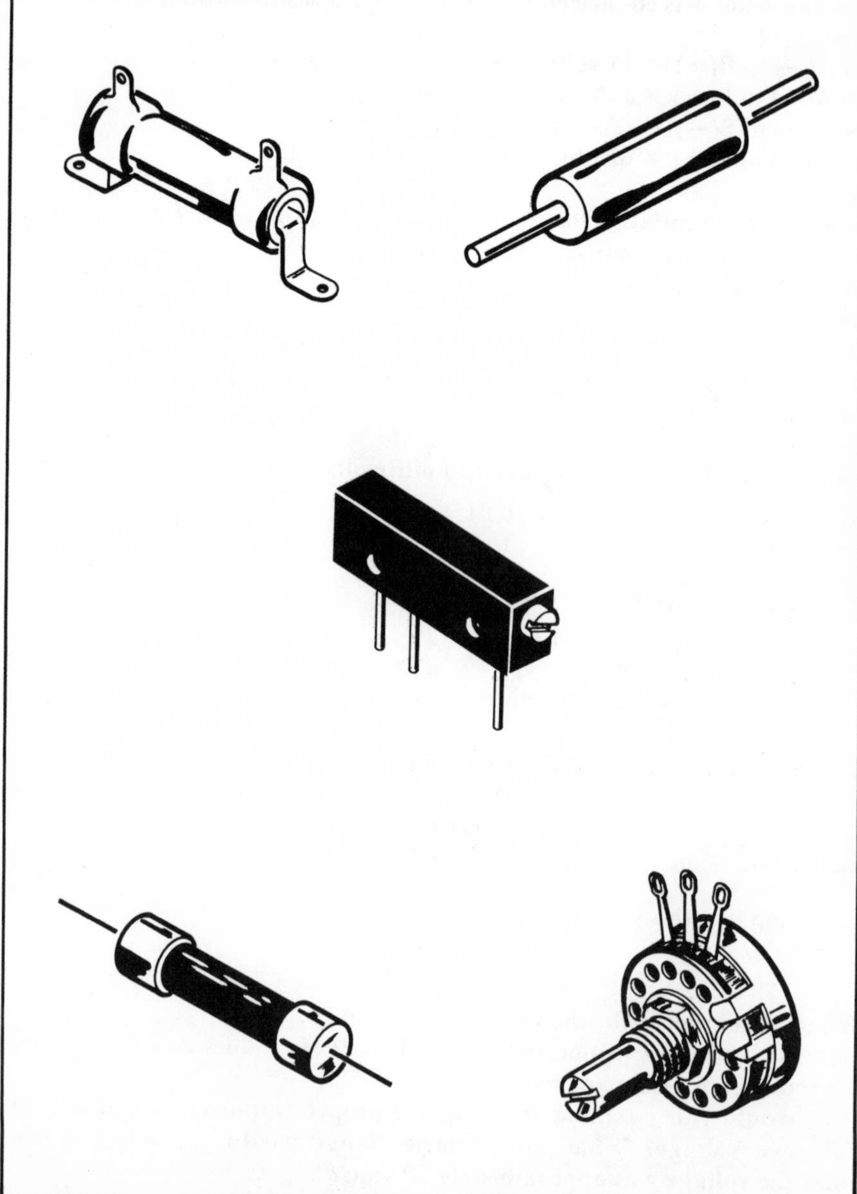

What Resistance Is

You know that an electric current is the movement of *free* electrons in a material, and that an electric current does not begin flowing all by itself because it needs a source of electric force to move the *free electrons* through the material. You have also found out that an electric current will not continue to flow if the source of electrical energy is removed. You can see from all this that there is something in a material that *resists* the flow of electric current—something that *holds onto* the *free* electrons and will not release them until force is applied.

As you learned earlier, the opposition to current flow is not the same for all materials. Current flow is the movement of *free* electrons through a material, and the *number* of *free* electrons in a material determines its *opposition* to current flow. Atoms of some materials give up their outer electrons easily. Such materials, called *conductors*, offer little opposition to current flow. Other materials hold onto their outer electrons. These materials, called *insulators*, offer considerable opposition to current flow. Every material has some opposition to current flow, whether large or small, and this opposition is called *resistance*.

What Resistance Is (continued)

Let's assume that we have a source of constant electric force (voltage). If this is the case, then the more opposition that we have to current flow (resistance), the smaller will be the number of electrons flowing (current) through the material. Also, the *converse* is true. That is, for a constant voltage, the *smaller* the resistance (less opposition), the *greater* the current flow (more electrons).

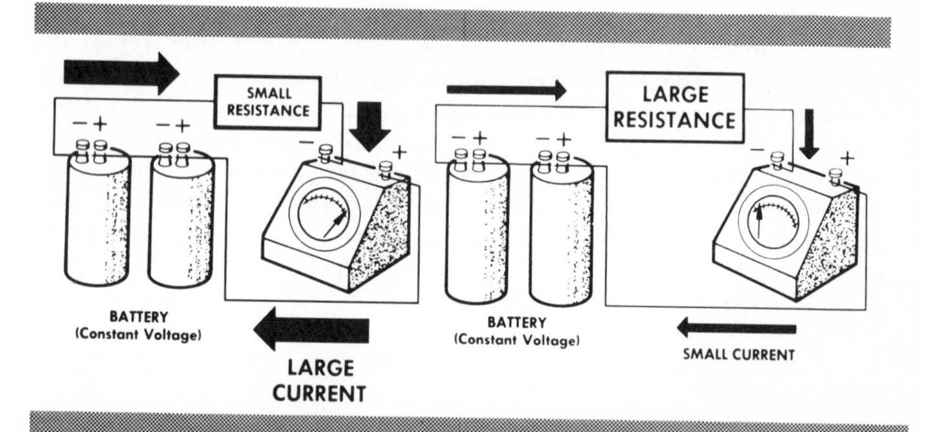

In fact, if you were to measure the current in a circuit such as the one above with one resistance element in it, and then were to *double* the amount of resistance in the circuit by putting another identical resistance element in series with it, you would find that the current would be equal to *one half* the original value. Thus, under conditions of constant voltage (electrical force), the *current flow* is *proportional* to the amount of resistance in the circuit.

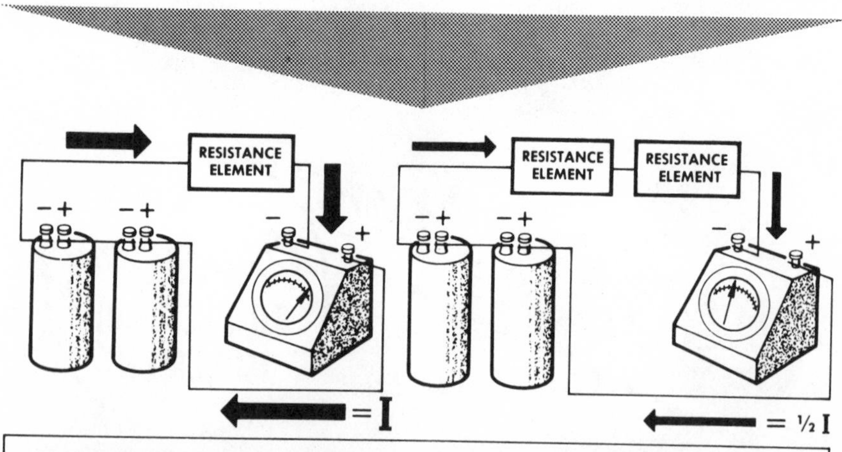

AT CONSTANT VOLTAGE, **DOUBLE** THE RESISTANCE = **ONE-HALF** THE CURRENT

What Resistance Is (continued)

Let's see now what happens when you hold the resistance constant and *vary* the voltage (electric force). You would expect that *increasing* the voltage (increased electric force) would permit the flow of *more* electrons (current) against the opposing effect of the resistance. This is exactly what happens.

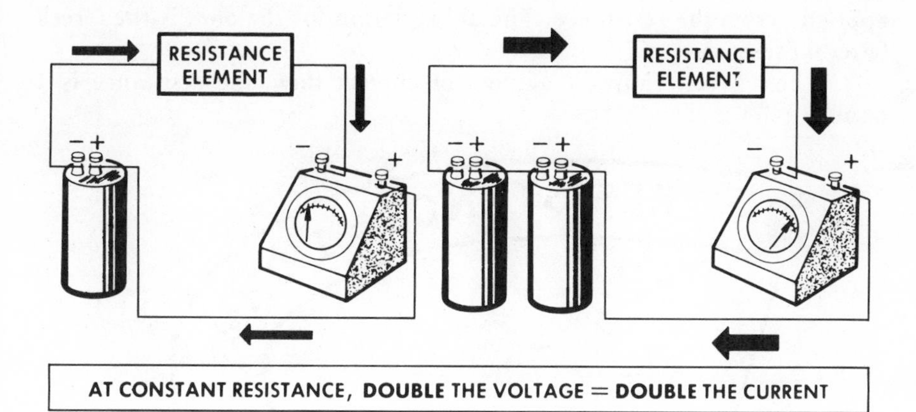

AT CONSTANT RESISTANCE, **DOUBLE** THE VOLTAGE = **DOUBLE** THE CURRENT

The above principles relating current flow to the amount of resistance in a circuit, or the voltage in a circuit, are of extreme importance and basic to the study of electricity. We will study these principles in much more detail later and use them throughout our study of electricity.

Although all conductors have resistance, you will have many occasions when you want to put in a *specific* amount of resistance in a circuit. Devices having known values of resistance are called *resistors*, designated with the letter "R," and are shown in circuit diagrams by the schematic symbol below.

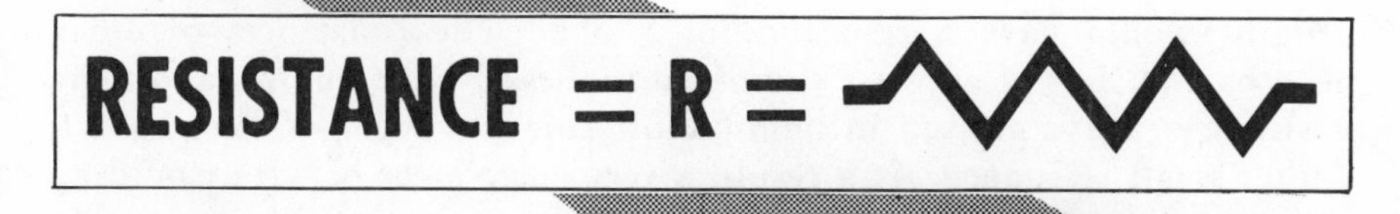

Units of Resistance

To measure current, the ampere is used as a unit of measure. To measure voltage, the volt is used. These units are necessary in order to compare different currents and different voltages. In the same manner, a unit of measure is needed to compare the resistance of different conductors. The basic unit of resistance is the *ohm*, equal to that resistance which will allow exactly *1 ampere* of *current* to flow when *1 volt* of *emf* is applied across the resistance. The abbreviation for the ohm is the Greek letter Ω (omega).

When 1 volt causes 1 ampere of current flow, the resistance is 1 ohm.

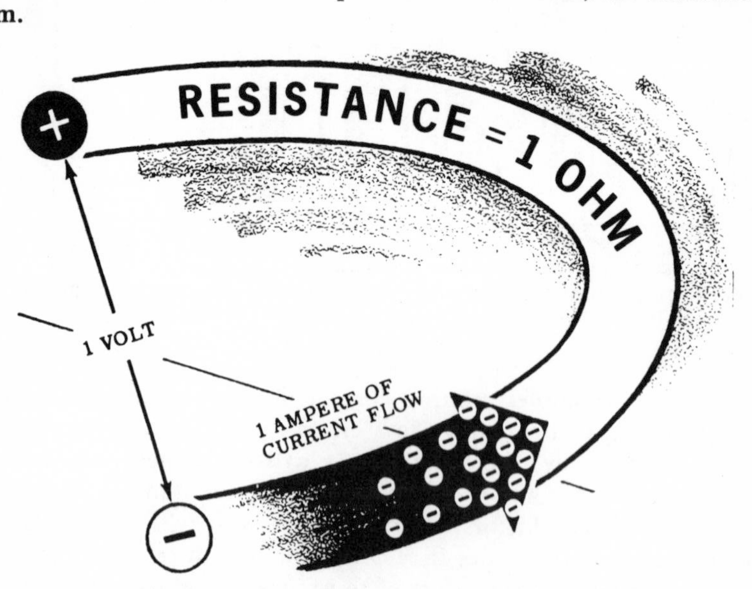

Suppose you connect a copper wire across a voltage source of 1 volt and adjust the length of the wire until the current flow through the wire is exactly 1 ampere. The resistance of the length of copper wire then is exactly 1 ohm. If you were to use wire of any other materials—iron, silver, etc.—you would find that the wire length and size would not be the same as that for copper. However, in each case you could find a length of the wire which would allow exactly 1 ampere of current to flow when connected across a 1-volt voltage source, and each of these lengths would have a resistance of 1 ohm. The resistances of other lengths and sizes of wire are compared to these 1-ohm lengths, and their resistances are expressed in ohms. Most common types of wire have a rather small resistance. As a result, a very large piece of wire would be necessary to get a large resistor. To get large resistors with reasonable size, special wires called *resistance wires* are used, or as is common in electronic circuits, resistors are made of a molded material made up from carbon and clay. In this way, it is possible to get a large resistance in a small space.

Units of Resistance (continued)

Although you will often find that resistance values are given in ohms, you will also find that in many cases large values of resistance will be used or indicated. In addition, on some occasions you will find it necessary to use small fractional values of the ohm. The previously learned prefixes—*micro*, *milli*, and *kilo*—that you have used with voltage and current are also used in the *same* way with resistance. In addition, we use another prefix *meg*, which, when put in front of *ohm*, is equal to 1,000,000 ohms; that is, 1 megohm is equal to 1,000,000 ohms.

Units of resistance are converted in the same manner as units of current or voltage. You will have to learn some new abbreviations, however, since K is often used to indicate *kilohms* and M or *meg* is often used to indicate *megohms*. Thus, 10 kilohms would be shown as 10 K and 3.3 megohms would be shown as 3.3 M or 3.3 meg.

CONVERTING RESISTANCE UNITS

MICROHMS TO OHMS

Move the decimal point 6 places to the left

35,000 microhms = 0.035 ohm

OHMS TO MICROHMS

Move the decimal point 6 places to the right

3.6 ohms = 3,600,000 microhms

MILLIOHMS TO OHMS

Move the decimal point 3 places to the left

2,700 milliohms = 2.7 ohms

OHMS TO MILLIOHMS

Move the decimal point 3 places to the right

0.68 ohms = 680 milliohms

KILOHMS TO OHMS

Move the decimal point 3 places to the right

6.2K = 6,200 ohms

OHMS TO KILOHMS

Move the decimal point 3 places to the left

47,000 ohms = 47 K

MEGOHMS TO OHMS

Move the decimal place 6 places to the right

2.7 Meg = 2,700,000 ohms

OHMS TO MEGOHMS

Move the decimal point 6 places to the left

620,000 ohms = 0.62 Meg

Factors Controlling Resistance

All materials have some resistance. In some cases this is *desirable*, as, for example, where you want to limit current flow deliberately and, thus, use components called *resistors* made up of materials chosen for their resistance properties. In other cases, resistance is an *undesirable* property and you want to keep it at a *minimum* as, for example, in the case where you want to deliver a large current to a load and do not want the current limited by the conductors. You will learn more about the undesirable, or desirable, resistance of conductors later when you study electric power. Now we will look at the factors that *control* resistance in a material.

The resistance of any object, such as a wire conductor, depends on four factors—the *material* from which it is made, the *length* of that material, the *cross-sectional area* of the material, and, finally, the *temperature* of the material.

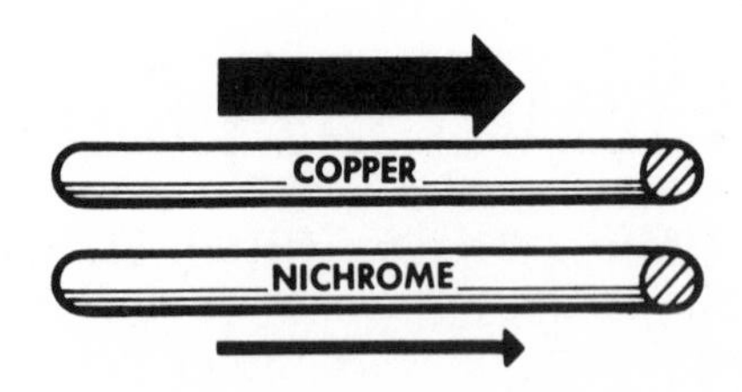

1. THE MATERIAL—Different materials have different resistances. Some, such as silver and copper, have a low resistance, while others, such as iron or nichrome (a special alloy of nickel, chromium and iron), have a higher resistance. Many resistors, such as those used in electronic circuits, are made of a molded mixture of carbon and clay.

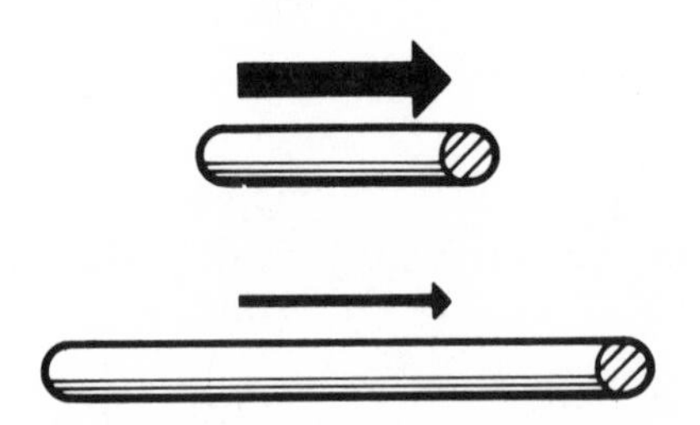

2. THE LENGTH—For a given material that has a constant cross-sectional area, the total resistance is *proportional* to the length. If a given length of the material has a resistance of 3 ohms, then twice that length will have a resistance of 6 ohms, a length of 3 times will have a resistance of 9 ohms, etc.

Factors in Controlling Resistance (continued)

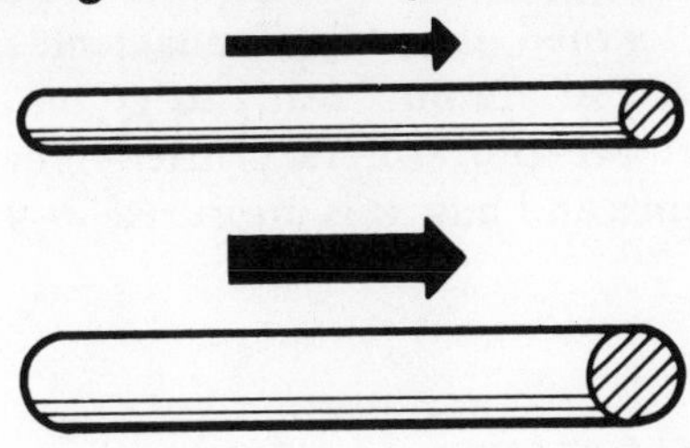

3. THE CROSS-SECTIONAL AREA—Current flow can be compared to the flow of water in a pipe. We know that if we make a pipe bigger (an increase in cross-sectional area), more water will flow even though the pressure is the same. There is a similar situation with respect to a conductor, in that the resistance *decreases* as the cross section *increases*. If we *double* the cross section of a material at constant length, the resistance will be *halved*. If we make the cross section one half, the resistance will be doubled.

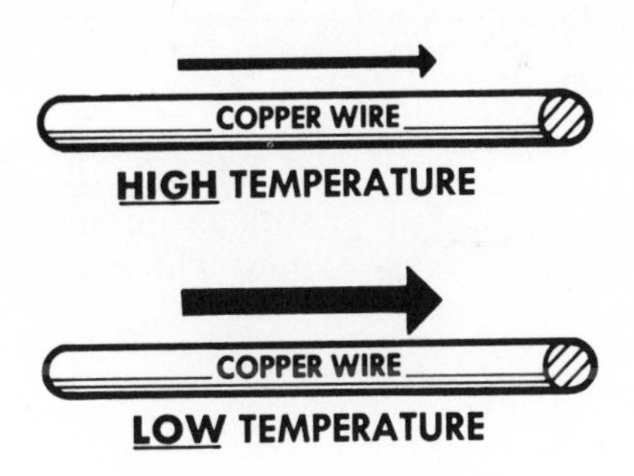

4. THE TEMPERATURE—Although temperature effects are generally small compared to the effects of material, length, and cross section, they can be important, particularly when we want to keep a resistance at a fixed value and the temperature is not constant. Generally, in metals, the resistance increases as the temperature increases. This is basically caused by the fact that the heat energy makes the free electrons in the material bounce around readily; and it is more difficult to get these electrons to flow along from atom to atom in an orderly way that we call current flow. In a few materials, such as carbon, the resistance decreases as the temperature increases.

Review of Resistance

You have now learned about the fundamental qualities of *voltage*, *current*, and *resistance*, and are now ready to go on and see how electric circuits work. Before we proceed, let's briefly review what you have learned about resistance and how it is measured.

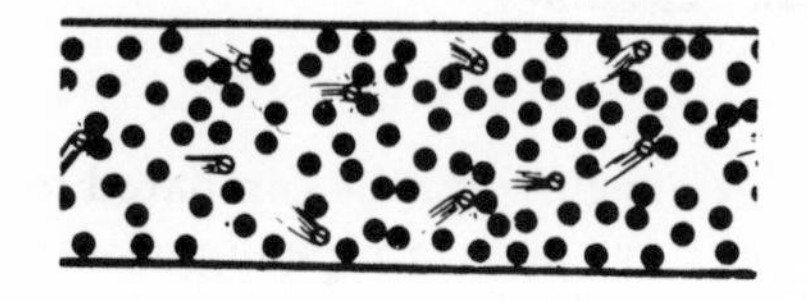

1. RESISTANCE—Opposition offered by a material to the flow of current.

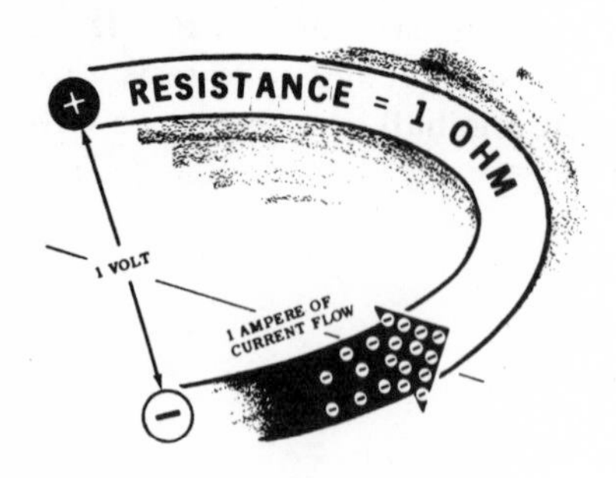

2. OHM—Basic unit of resistance measure equal to that resistance which allows 1 ampere of current to flow when an emf of 1 volt is applied across the resistance. The symbol for the ohm is Ω.

3. RESISTOR—Device having resistance used to control current flow. The symbol for a resistor is *R*.

4. OHMMETER—Meter used to measure resistance directly.

1 K = 1000Ω

5. KILOHM—One kilohm equals 1,000 ohms.

1 MegΩ = 1,000,000Ω

6. MEGOHM—One megohm is equal to 1,000,000 ohms.

Self-Test—Review Questions

1. Define what resistance is. What is a resistor? What is the symbol used to designate a resistor?
2. In a circuit with constant voltage, what happens to the current when the resistance is doubled? Halved? Tripled?
3. In a circuit with constant resistance, what happens to the current when the voltage is doubled? Halved? Quadrupled? Tripled?
4. Define the unit of resistance. What symbol is used to designate it?
5. What factors determine the resistance of a resistor? Give examples of their effect.

Calculate the following conversions using appropriate symbols where applicable:

6. Convert to ohms

   6.2 K
   6.2 M
   270 milliohms
   3.3 K
   9.1 kilohms
   4.7 megohms

7. Convert to kilohms

   4,700 ohms
   8.2 megohms
   100,000 ohms
   0.1 megohms
   0.39 megohms
   24,000 ohms

8. Convert to megohms

   1,000 kilohms
   120,000 ohms
   82,000 ohms
   68 K
   470,000 ohms
   330 K

9. Draw a schematic diagram of a resistor. What is the symbol used to designate resistance?
10. What are the four factors that affect resistance in a conductor? How does the resistance vary as these change?

Review of Current (I), Voltage (E), and Resistance (R)

As a conclusion to your study of electricity in action, you should consider again what you have found out about current, voltage, and resistance.

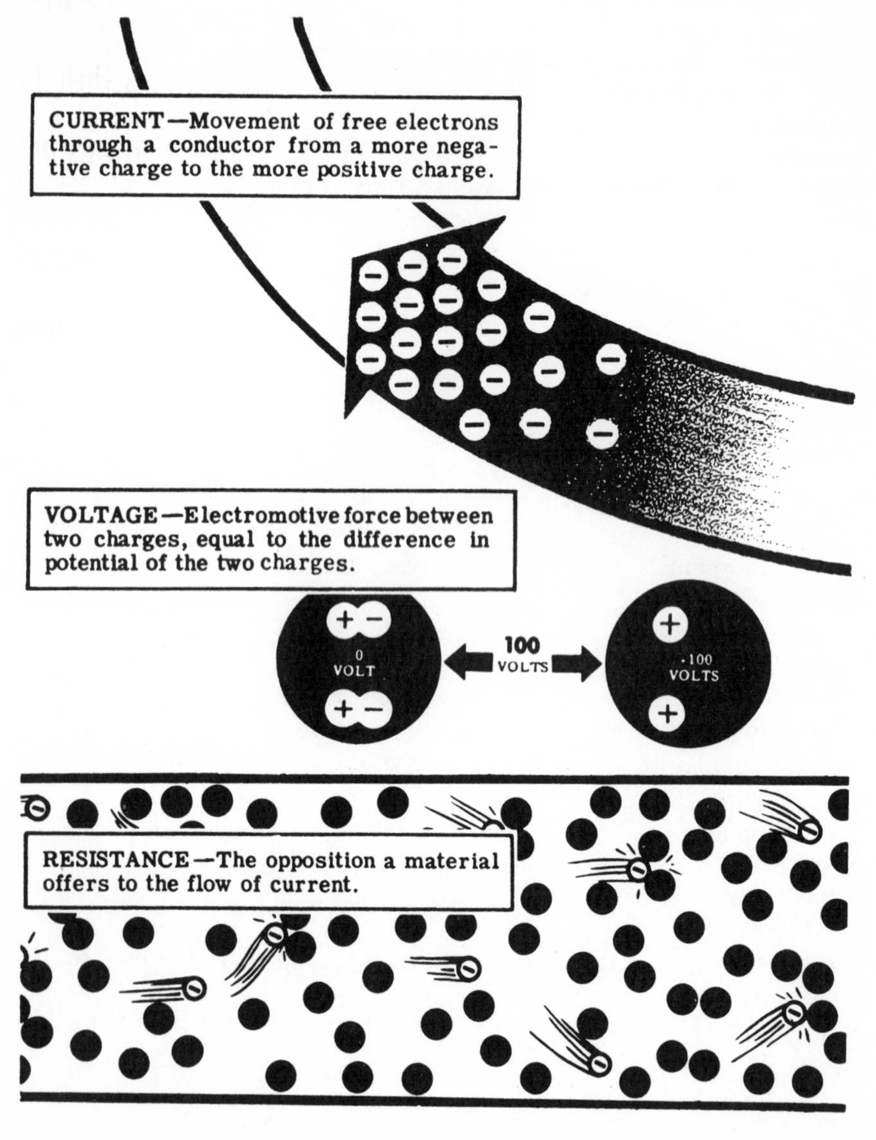

Particularly, you should recall the relationships between *current*, *voltage*, and *resistance*. Current flow is caused by the voltage between two points and is limited by the resistance between the points. In continuing your study of electricity, you will next find out about electric circuits and how they use current, voltage, and resistance.

The Relationship of Current, Voltage, and Resistance

As we said initially, the study of electricity is the study of the *effects* of the flow of current and the *control* of the flow of current.

Voltage, as you know, is the amount of electromotive force (emf) that is applied across a load (resistance) in order to make an electron current flow through the resistance. As you have learned, the *greater* the voltage you apply across a resistance, the *greater* the current flow. Similarly, the *lower* the voltage you apply, the *smaller* will be the current flow.

Resistance, as you also know, is the effect that impedes the flow of electrons. If you *increase* the resistance of the load across which a constant voltage is applied, *less* current will flow. Similarly, the *lower* you make the resistance, the *greater* will be the current flow.

This relationship between voltage, resistance, and current was studied by the German mathematician George Simon Ohm. His description, now know as *Ohm's Law*, says that current varies directly with the voltage and inversely with the resistance. The mathematical analysis of the law is of no concern to you at present, but you will learn about it when you get into Volume 2. Ohm's Law is a *basic tool* for all who work with electric circuits in any shape or form.

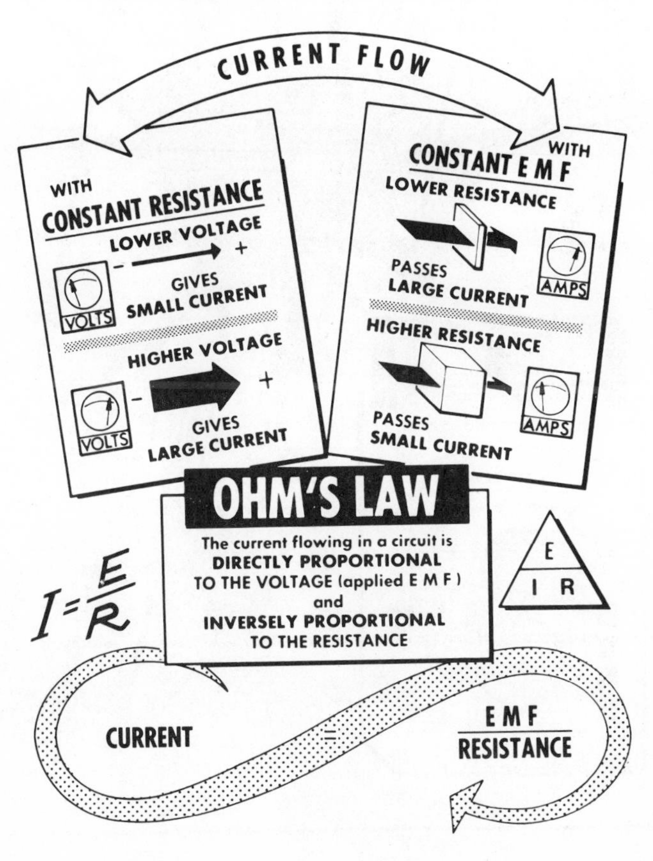

Learning Objectives—Next Volume

Overview—Now that you have learned some fundamentals of electricity, and know about current, voltage and resistance, you are ready to study the electric circuits in Volume 2. Sources and loads must be connected together to function and when connected together, are called *electric circuits*. What electric circuits are and how they are connected and behave is the most important part of the study of electricity.

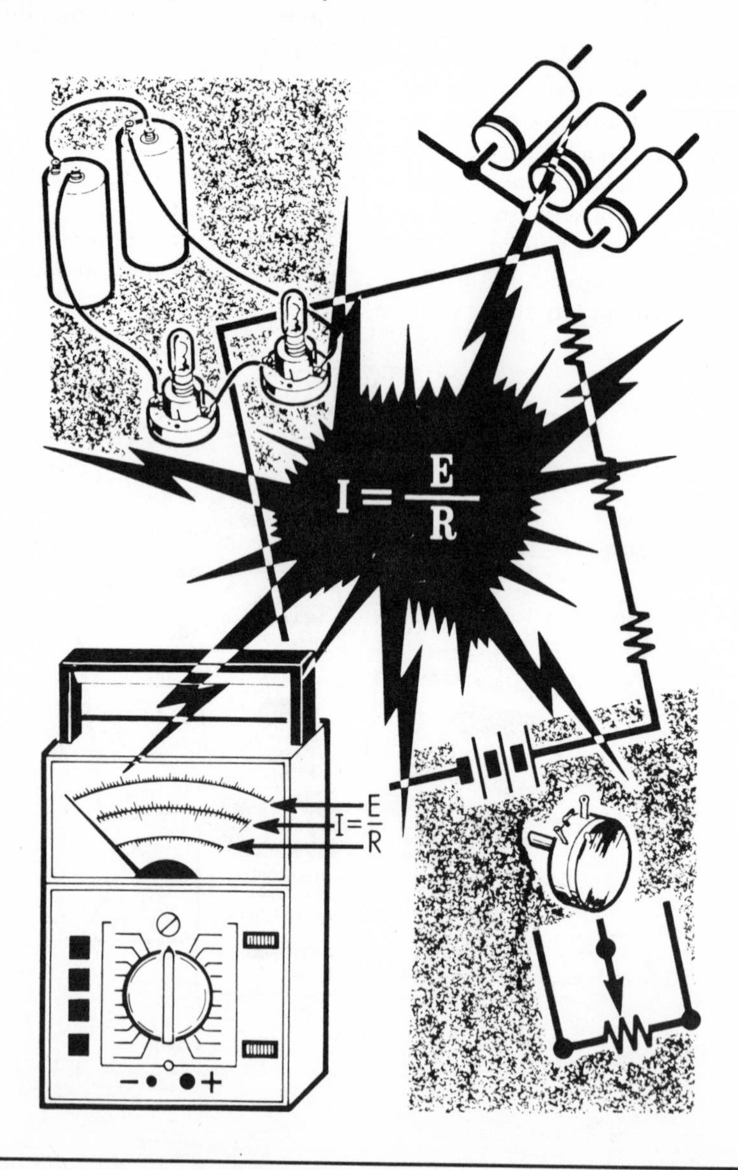

TABLE OF THE ELEMENTS

Atomic Number	Name of Element	Symbol of Element	Atomic Weight	Atomic Number	Name of Element	Symbol of Element	Atomic Weight
1	Hydrogen	H	1	53	Iodine	I	127
2	Helium	He	4	54	Xenon	Xe	131
3	Lithium	Li	7	55	Cesium	Cs	133
4	Beryllium	Be	9	56	Barium	Ba	137
5	Boron	B	11	57	Lanthanum	La	139
6	Carbon	C	12	58	Cerium	Ce	140
7	Nitrogen	N	14	59	Praseodymium	Pr	141
8	Oxygen	O	16	60	Neodymium	Nd	144
9	Fluorine	F	19	61	Promethium	Pm	147
10	Neon	Ne	20	62	Samarium	Sm	150
11	Sodium	Na	22	63	Europium	Eu	152
12	Magnesium	Mg	24	64	Gadolinium	Gd	157
13	Aluminum	Al	27	65	Terbium	Tb	159
14	Silicon	Si	28	66	Dysprosium	Dy	162
15	Phosphorus	P	31	67	Hilmium	Ho	165
16	Sulfur	S	32	68	Erbium	Er	167
17	Chlorine	Cl	35	69	Thulium	Tm	169
18	Argon	A	39	70	Ytterbium	Yb	173
19	Potassium	K	39	71	Lutecium	Lu	175
20	Calcium	Ca	40	72	Hafnium	Hf	179
21	Scandium	Sc	45	73	Tantalum	Ta	181
22	Titanium	Ti	48	74	Tungsten	W	184
23	Vanadium	V	51	75	Rhenium	Re	186
24	Chromium	Cr	52	76	Osmium	Os	190
25	Manganese	Mn	55	77	Iridium	Ir	193
26	Iron	Fe	56	78	Platinum	Pt	195
27	Cobalt	Co	59	79	Gold	Au	197
28	Nickel	Ni	59	80	Mercury	Hg	201
29	Copper	Cu	64	81	Thallium	Tl	204
30	Zinc	Zn	65	82	Lead	Pb	207
31	Gallium	Ga	70	83	Bismuth	Bi	209
32	Germanium	Ge	73	84	Polonium	Po	210
33	Arsenic	As	75	85	Astatine	At	211
34	Selenium	Se	79	86	Radon	Rn	222
35	Bromine	Br	80	87	Francium	Fr	223
36	Krypton	Kr	84	88	Radium	Ra	226
37	Rubidium	Rb	85	89	Actinium	Ac	227
38	Strontium	Sr	88	90	Thorium	Th	232
39	Yttrium	Y	89	91	Protactinium	Pa	231
40	Zirconium	Zr	91	92	Uranium	U	238
41	Columbium	Cb	93	93	Neptunium	Np	239
42	Molybdenum	Mo	96	94	Plutonium	Pu	239
43	Technetium	Tc	99	95	Americium	Am	241
44	Ruthenium	Ru	102	96	Curium	Cm	242
45	Rhodium	Rh	103	97	Berkelium	Bk	245
46	Palladium	Pd	107	98	Californium	Cf	246
47	Silver	Ag	108	99	Einsteinium	E	253
48	Cadmium	Cd	112	100	Fermium	Fm	256
49	Indium	In	115	101	Mendelevium	Mv	256
50	Tin	Sn	119	102	Nobelium	No	254
51	Antimony	Sb	122	103	Lawrencium	Lw	257
52	Tellurium	Te	128	104 & 105	Under Study		

Note: Elements 1 through 92 occur normally in nature. Elements 93 and above are those discovered by man as a result of transmutation.

INDEX TO VOL. 1

(Note: A cumulative index covering all five volumes in this series will be found at the end of Volume 5.)